VOLUME ONE HUNDRED AND FORTY TWO

ADVANCES IN
CANCER RESEARCH

VOLUME ONE HUNDRED AND FORTY TWO

Advances in
CANCER RESEARCH

Edited by

KENNETH D. TEW
*Department of Cell and Molecular Pharmacology
and Experimental Therapeutics,
Medical University of South Carolina, Charleston,
SC, United States*

PAUL B. FISHER
*Department of Human and Molecular Genetics;
VCU Institute of Molecular Medicine;
VCU Massey Cancer Center, Virginia Commonwealth
University School of Medicine, Richmond,
VA, United States*

Academic Press is an imprint of Elsevier
50 Hampshire Street, 5th Floor, Cambridge, MA 02139, United States
525 B Street, Suite 1650, San Diego, CA 92101, United States
The Boulevard, Langford Lane, Kidlington, Oxford OX5 1GB, United Kingdom
125 London Wall, London, EC2Y 5AS, United Kingdom

First edition 2019

Copyright © 2019 Elsevier Inc. All rights reserved.

No part of this publication may be reproduced or transmitted in any form or by any means, electronic or mechanical, including photocopying, recording, or any information storage and retrieval system, without permission in writing from the publisher. Details on how to seek permission, further information about the Publisher's permissions policies and our arrangements with organizations such as the Copyright Clearance Center and the Copyright Licensing Agency, can be found at our website: www.elsevier.com/permissions.

This book and the individual contributions contained in it are protected under copyright by the Publisher (other than as may be noted herein).

Notices
Knowledge and best practice in this field are constantly changing. As new research and experience broaden our understanding, changes in research methods, professional practices, or medical treatment may become necessary.

Practitioners and researchers must always rely on their own experience and knowledge in evaluating and using any information, methods, compounds, or experiments described herein. In using such information or methods they should be mindful of their own safety and the safety of others, including parties for whom they have a professional responsibility.

To the fullest extent of the law, neither the Publisher nor the authors, contributors, or editors, assume any liability for any injury and/or damage to persons or property as a matter of products liability, negligence or otherwise, or from any use or operation of any methods, products, instructions, or ideas contained in the material herein.

ISBN: 978-0-12-817153-0
ISSN: 0065-230X

For information on all Academic Press publications
visit our website at https://www.elsevier.com/books-and-journals

Publisher: Zoe Kruze
Acquisition Editor: Fiona Pattison
Editorial Project Manager: Shellie Bryant
Production Project Manager: Vijayaraj Purushothaman
Cover Designer: Greg Harris

Typeset by SPi Global, India

Contents

Contributors   ix

1. **Highly variant DNA methylation in normal tissues identifies a distinct subclass of cancer patients**   1
   Jayashri Ghosh, Bryant Schultz, Christos Coutifaris, and Carmen Sapienza

   1. Introduction   2
   2. Identifying OMP   2
   3. OMP patients in TCGA   7
   4. Do OMP patients have CIMP tumors?   10
   5. Are there pan-cancer OMP CpGs?   12
   6. Do OMP patients constitute a distinct subgroup of cancer patients?   13
   7. Do OMP patients have distinct tumor epigenomes?   16
   8. Are OMP individuals prone to adverse outcomes?   17
   9. What are potential causes of OMP?   17
   10. Conclusion and future directions   19

   Conflict of interest   19
   References   20

2. **Bittersweet tumor development and progression: Emerging roles of epithelial plasticity glycosylations**   23
   Ryan M. Phillips, Christine Lam, Hailun Wang, and Phuoc T. Tran

   1. Epithelial plasticity—Basics and implications in cancer development, progression and treatment resistance   24
   2. Cancer metabolism—Metabolic adaptations in cancer   27
   3. Glycobiology—An introduction to protein glycosylation and relevance to cancer   30
   4. The hexosamine biosynthesis pathway (HBP)—An emerging metabolic player in cancer   32
   5. Epithelial plasticity and glycosylations in cancer   35
   6. Associations between EMT and O-GlcNAcylation   45
   7. The EMT-HBP-O-GlcNAcylation axis—An important new pathway to promote the neoplastic phenotype   48
   8. Conclusion and future directions   54

   Acknowledgments   55
   Conflict of interest   55
   References   55

3. **The second genome: Effects of the mitochondrial genome on cancer progression**   **63**
Adam D. Scheid, Thomas C. Beadnell, and Danny R. Welch

1. Mitochondrial evolution and genetic variation	65
2. Mitochondria and cancer	67
3. Studying direct mtDNA contributions to disease	75
4. Concluding remarks and remaining questions	90
Acknowledgments	91
Conflict of interest	91
References	91

4. **Pathways- and epigenetic-based assessment of relative immune infiltration in various types of solid tumors**   **107**
Manny D. Bacolod, Francis Barany, Karsten Pilones, Paul B. Fisher, and Romulo J. de Castro

1. Introduction	108
2. Methodology	110
3. Results	114
4. Discussion	133
5. Expert opinion	138
Acknowledgments	139
References	139

5. **HVEM network signaling in cancer**   **145**
John R. Šedý and Parham Ramezani-Rad

1. Introduction	148
2. HVEM network interactions in the immune microenvironment	149
3. Genetic lesions in *TNFRSF14* in lymphoid cancers	151
4. Genetic lesions in *TNFRSF14* in non-lymphoid cancers	158
5. HVEM functions within the tumor microenvironment	160
6. Therapeutic targeting of the HVEM network in lymphoma and other tumors	169
7. Concluding remarks	172
Funding	173
References	173

6. Pharmacology of ME-344, a novel cytotoxic isoflavone — 187
Leilei Zhang, Jie Zhang, Zhiwei Ye, Danyelle M. Townsend, and Kenneth D. Tew

1. Introduction — 188
2. Mitochondria as drug target organelles in cancer — 190
3. Cell components targeted by ME-344 — 191
4. Preclinical investigational new drug (IND) enabling studies for ME-344 — 198
5. Clinical studies — 200
6. Conclusions and future perspectives — 203

Acknowledgments — 203
Conflict of interest — 204
References — 204

Contributors

Manny D. Bacolod
Department of Microbiology and Immunology, Weill Cornell Medicine, New York, NY, United States

Francis Barany
Department of Microbiology and Immunology, Weill Cornell Medicine, New York, NY, United States

Thomas C. Beadnell
Department of Cancer Biology, The University of Kansas Medical Center, and The University of Kansas Cancer Center, Kansas City, KS, United States

Christos Coutifaris
Department of Obstetrics & Gynecology, University of Pennsylvania School of Medicine, Philadelphia, PA, United States

Romulo J. de Castro
3R Biosystems, Long Beach, CA, United States

Paul B. Fisher
Department of Human and Molecular Genetics, VCU Institute of Molecular Medicine, VCU Massey Cancer Center, Virginia Commonwealth University, School of Medicine, Richmond, VA, United States

Jayashri Ghosh
Fels Institute of Cancer Research and Molecular Biology, Lewis Katz School of Medicine, Temple University, Philadelphia, PA, United States

Christine Lam
Department of Radiation Oncology and Molecular Radiation Sciences, Sidney Kimmel Comprehensive Cancer Center; Program in Cellular and Molecular Medicine, Johns Hopkins University School of Medicine, Baltimore, MD, United States

Ryan M. Phillips
Department of Radiation Oncology and Molecular Radiation Sciences, Sidney Kimmel Comprehensive Cancer Center, Johns Hopkins University School of Medicine, Baltimore, MD, United States

Karsten Pilones
Department of Radiation Oncology, Weill Cornell Medicine, New York, NY, United States

Parham Ramezani-Rad
Infectious and Inflammatory Disease Center, NCI-Designated Cancer Center, Sanford Burnham Prebys Medical Discovery Institute, La Jolla, CA, United States

Carmen Sapienza
Fels Institute of Cancer Research and Molecular Biology; Department of Pathology and Laboratory Medicine, Lewis Katz School of Medicine, Temple University, Philadelphia, PA, United States

Adam D. Scheid
Department of Cancer Biology, The University of Kansas Medical Center, and The University of Kansas Cancer Center, Kansas City, KS, United States

Bryant Schultz
Fels Institute of Cancer Research and Molecular Biology, Lewis Katz School of Medicine, Temple University, Philadelphia, PA, United States

John R. Šedý
Infectious and Inflammatory Disease Center, NCI-Designated Cancer Center, Sanford Burnham Prebys Medical Discovery Institute, La Jolla, CA, United States

Kenneth D. Tew
Department of Cell and Molecular Pharmacology and Experimental Therapeutics, Medical University of South Carolina, Charleston, SC, United States

Danyelle M. Townsend
Drug Discovery and Biomedical Sciences, Medical University of South Carolina, Charleston, SC, United States

Phuoc T. Tran
Department of Radiation Oncology and Molecular Radiation Sciences, Sidney Kimmel Comprehensive Cancer Center; Program in Cellular and Molecular Medicine; Department of Urology, James Buchanan Brady Urological Institute; Department of Oncology, Sidney Kimmel Comprehensive Cancer Center, Johns Hopkins University School of Medicine, Baltimore, MD, United States

Hailun Wang
Department of Radiation Oncology and Molecular Radiation Sciences, Sidney Kimmel Comprehensive Cancer Center, Johns Hopkins University School of Medicine, Baltimore, MD, United States

Danny R. Welch
Department of Cancer Biology, The University of Kansas Medical Center, and The University of Kansas Cancer Center, Kansas City, KS, United States

Zhiwei Ye
Department of Cell and Molecular Pharmacology and Experimental Therapeutics, Medical University of South Carolina, Charleston, SC, United States

Jie Zhang
Department of Cell and Molecular Pharmacology and Experimental Therapeutics, Medical University of South Carolina, Charleston, SC, United States

Leilei Zhang
Department of Cell and Molecular Pharmacology and Experimental Therapeutics, Medical University of South Carolina, Charleston, SC, United States

CHAPTER ONE

Highly variant DNA methylation in normal tissues identifies a distinct subclass of cancer patients

Jayashri Ghosh[a], Bryant Schultz[a], Christos Coutifaris[b], Carmen Sapienza[a,c,]*

[a]Fels Institute of Cancer Research and Molecular Biology, Lewis Katz School of Medicine, Temple University, Philadelphia, PA, United States
[b]Department of Obstetrics & Gynecology, University of Pennsylvania School of Medicine, Philadelphia, PA, United States
[c]Department of Pathology and Laboratory Medicine, Lewis Katz School of Medicine, Temple University, Philadelphia, PA, United States
*Corresponding author: e-mail address: sapienza@temple.edu

Contents

1. Introduction 2
2. Identifying OMP 2
   2.1 How OMP differs from CIMP 5
   2.2 Why identify OMP in cancer patients? 7
3. OMP patients in TCGA 7
4. Do OMP patients have CIMP tumors? 10
5. Are there pan-cancer OMP CpGs? 12
6. Do OMP patients constitute a distinct subgroup of cancer patients? 13
7. Do OMP patients have distinct tumor epigenomes? 16
8. Are OMP individuals prone to adverse outcomes? 17
9. What are potential causes of OMP? 17
10. Conclusion and future directions 19
Conflict of interest 19
References 20

Abstract

The "CpG Island Methylator Phenotype" (CIMP) has been found to be a useful concept in stratifying several types of human cancer into molecularly and clinically distinguishable subgroups. We have identified an additional epigenetic stratification category, the "Outlier Methylation Phenotype" (OMP). Whereas CIMP is defined on the basis of hyper-methylation in *tumor* genomes, OMP is defined on the basis of highly variant (either or both hyper- and hypo-methylation) methylation at many sites in *normal* tissues. OMP was identified and defined, originally, as being more common among low birth weight individuals conceived in vitro but we have also identified OMP individuals among colon cancer patients profiled by us, as well as multiple types of cancer patients in the TCGA database. The cause(s) of OMP are unknown, as is whether these individuals identify a clinically useful subgroup of patients, but both the causes of, and potential consequences to, this epigenetically distinct group are of great interest.

1. Introduction

We begin this review on "highly variant DNA methylation in normal tissues" with a sense of bemusement. Why should one review the literature on a field with so little primary literature (Gaykalova et al., 2015; Ghosh, Mainigi, Coutifaris, & Sapienza, 2016; Li et al., 2015; Mamatjan et al., 2017; Teschendorff et al., 2012; Teschendorff, Gao, et al., 2016; Teschendorff, Jones, & Widschwendter, 2016)? However, the Editors felt that most Readers would be unaware of the topic and it was likely to be of interest, if for no other reason than to encourage the epigenetics community to analyze epigenome-wide DNA methylation data from a slightly different and broader perspective.

Our overall hypothesis is that much of human phenotypic variation, including risk of common diseases, has its roots in epigenetic variation. As with any type of human variation, examination of the extremes has often shed light on which factors have the largest effect on the phenotype in question. Our focus is on identifying individuals who have normal tissue DNA methylation levels that vary dramatically from the population mean at a significant fraction of all sites profiled (defined as the "Outlier Methylation Phenotype"—OMP) (Ghosh et al., 2016), determining whether these individuals constitute a distinct phenotypic or clinical group and whether this normal tissue OMP characteristic presages a particular genetic or epigenetic condition in tumor tissues.

In this review, we will first define OMP and explain how it differs from "CpG Island Methylator Phenotype" (CIMP) (Toyota et al., 1999). We also identify OMP individuals in the TCGA database and explore whether they constitute distinct subclasses of cancer patients as well as what might be some of the potential causes of OMP.

2. Identifying OMP

While there are a variety of statistical procedures by which one might identify individuals who are at the extremes of population variation in DNA methylation (Teschendorff, Jones, & Widschwendter, 2016), i.e., so-called outliers, we have used a simple parametric procedure taken from our published work (Ghosh et al., 2016). In that particular instance, we were able to associate a molecular phenotype (OMP) with an environmental factor (in vitro fertilization) and an undesirable clinical outcome (low birth weight).

The binary classification of individuals as having "Outlier Methylation Phenotype" (or not) is accomplished using a two-step process. In the first step, each individual is classified as having/not having an "outlier" methylation level at every CpG sites interrogated. This process is illustrated in Fig. 1 (taken from Ghosh et al., 2016). Each symbol in Fig. 1A represents the methylation level of a single individual at a single CpG site in *GRB10* (designated cg24274319 on Illumina platforms). We used a common algorithm to designate "outliers" as any individual whose methylation level exceeded the interquartile range of the distribution of methylation levels of all individuals interrogated at that site by >50%. The eight individuals inside the circle in Fig. 1A display "outlier" levels of methylation at this particular CpG. The assignment of each individual as outlier/non-outlier is then made for every CpG interrogated (as in Figs. 1B–D) and the number of CpGs for which each individual is scored as an outlier is tabulated. For example, the individual denoted by a downward-pointing triangle in Fig. 1A–D is an outlier at all four CpG sites shown in the figure.

In the second step of the process, the distribution of individuals with respect to the total number of outlier CpGs for which each is an outlier is subjected to the same algorithm. For example, in Fig. 2 (again taken from Ghosh et al., 2016), the distribution of all low birth weight children with respect to the number of outlier CpGs has been stratified according to mode of conception. Those individuals designated by open circles or open squares are characterized as having OMP by virtue of having >50% more outlier CpGs than those individuals within the interquartile range.

One thing that is worth noting in the original example of how we defined OMP (Ghosh et al., 2016) is that the direction of an individual's outlier methylation at each CpG site interrogated, i.e., whether they were hyper- or hypomethylated, compared with the rest of the population, was not taken into consideration. Because our original work was focused on potential reasons for the moderate epidemiological associations between assisted reproduction and a number of undesirable outcomes (Ghosh et al., 2016; Goelz, Vogelstein, Hamilton, & Feinberg, 1985; Mainigi, Sapienza, Butts, & Coutifaris, 2016), we sought to associate a molecular phenotype

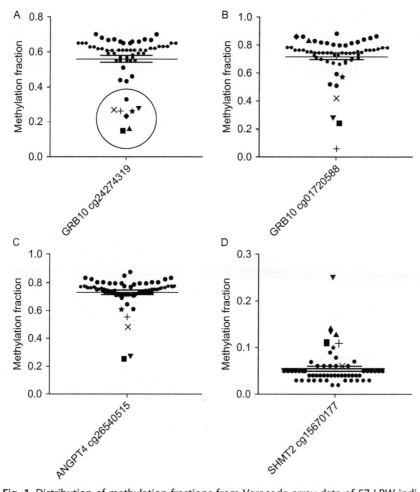

Fig. 1 Distribution of methylation fractions from Veracode array data of 57 LBW individuals in (A) *GRB10* cg24274319, (B) *GRB10* cg01720588, (C) *ANGPT4* cg26540515 and (D) *SHMT2* cg15670177. The symbols within the black circle in (A) represent individuals that are "outliers" at this CpG site. These individuals are represented by the same symbols (other than the closed circle) in (B–D). For example, the filled square symbol denotes the same individual in each graph. *Adapted from Ghosh, J., Mainigi, M., Coutifaris, C., & Sapienza, C. (2016). Outlier DNA methylation levels as an indicator of environmental exposure and risk of undesirable birth outcome.* Human Molecular Genetics. 25, 123–9.

(a strongly disrupted epigenome—characterized as OMP) with an environmental factor (assisted reproduction) and a particular undesirable outcome (low birth weight). In the case of cancer phenotypes, ignoring the direction of outlier methylation may obscure groups of OMP individuals who may be

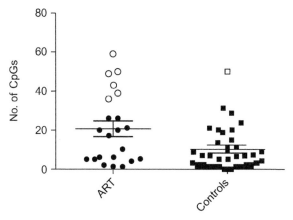

Fig. 2 The total number of CpGs in which each ART and Control individual is an outlier in low birth weight group. The open symbols indicate the "outliers" of each combined distribution. *Adapted from Ghosh, J., Mainigi, M., Coutifaris, C., & Sapienza, C. (2016). Outlier DNA methylation levels as an indicator of environmental exposure and risk of undesirable birth outcome.* Human Molecular Genetics. 25, 123–9.

stratified by common mechanisms; environmental exposures versus genetic predisposition, for example. Given the early observations on colon tumors that tumor DNA was globally hypomethylated (Goelz et al., 1985; Hansen et al., 2011; Sunami, de Maat, Vu, Turner, & Hoon, 2011), but that many CpG islands were hypermethylated (Berman et al., 2011; Toyota et al., 1999), we have stratified the OMP phenotype into hyper- and hypo-methylated subgroups in our analysis of cancer patients (below).

2.1 How OMP differs from CIMP

As mentioned earlier, CIMP (CpG Island Methylator Phenotype) is a molecular phenotype that has been used to stratify cancer patients into subgroups that appear to have some common molecular features and clinical significance. The original definition (Toyota et al., 1999) was based on hyper-methylation of a small number of CpG islands in colorectal cancer patients. However, the concept has evolved to the point where more than one type of CIMP has been described (Hinoue et al., 2012; Ogino et al., 2007; Shen et al., 2007), based on specific exposures or mutational signatures (Weisenberger et al., 2006).

Since the first report on CIMP (Toyota et al., 1999), a number of reports have characterized CIMP in colorectal cancers. Weisenberger et al. (2006)

identified CIMP positive clusters based on a five-gene panel that was characterized by mutations in *BRAF* and microsatellite instability (MSI-H). Later, several investigators also noted the presence of an intermediate CIMP subgroup (CIMP low or CIMPi), in addition to the traditional two-group classification (Hinoue et al., 2012; Ogino et al., 2007; Shen et al., 2007).

CIMP definition has evolved over the years with the identification of CIMP positive tumors in multiple cancer types (An et al., 2005; Arai et al., 2012; Bae et al., 2004; Maruyama et al., 2001, 2002; Miller, Sánchez-Vega, & Elnitski, 2016; Noushmehr et al., 2010; Strathdee et al., 2001; Suzuki et al., 2006; Tanemura et al., 2009; Toyota et al., 2001; Ueki et al., 2000; Whitcomb et al., 2003; Zhang et al., 2007). Investigators have also identified specific mutations associated with CIMP positive tumors in some cancer types (for example *IDH1* mutations in glioma, Noushmehr et al., 2010; *IDH1* and *IDH2* mutations in leukemia, Figueroa et al., 2010). The CIMP positive clusters are also based on their specificity to tumor types: C-CIMP for colon, G-CIMP for gliomas and B-CIMP for breast (Hughes et al., 2013). It appears that a partial CIMP signature is shared between these clusters across different cancer types, with many of them targeting the polycomb group (Fang et al., 2011). However, this overlap is not complete and CIMP definitions across different cancer types display heterogeneity.

Identification of CIMP positive or CIMP high tumors have been associated with poor prognosis or survival and in some studies have been linked to gene mutations, but different investigators have used different gene panels to identify CIMP positive tumors. It is noteworthy that after nearly 20 years of CIMP discovery, there is no clear consensus for the ideal CIMP marker panel for each tumor type or for a pan-cancer panel.

The biggest distinction between CIMP and OMP is that CIMP is based on hypermethylation of CpG sites *in tumor tissue*, while OMP is defined as highly variant methylation (either hypo-, hyper- or both hypo- and hypermethylation) *in normal tissue*. The major questions surrounding OMP are: (1) whether having a strongly disrupted epigenome, as evidenced by DNA methylation levels that are far from the population mean at many sites, predisposes one to cancers in comparison with the rest of the population; (2) whether the cancers of OMP individuals are characterized by additional or particular genetic or epigenetic events in tumor tissue (i.e., are OMP individuals more likely to develop CIMP positive tumors?), and (3) whether cancer patients with OMP constitute one or more distinct clinical groups.

2.2 Why identify OMP in cancer patients?

The notion that being an OMP individual may define the epigenomic consequences of an environmental exposure is not limited to the particular exposure of Assisted Reproductive Technologies (ART) (Ghosh et al., 2016). Moreover, the concept that a highly disrupted epigenome, as identified by OMP, is associated with adverse outcomes has gained credence in several disciplines. We believe our own study (Ghosh et al., 2016), associating this molecular phenotype with both a clinical procedure (ART) and an undesirable outcome (low birth weight) was the first example. More recently, Teschendorff et al. have demonstrated that the normal tissues of many breast cancer patients have large numbers of outlier epigenetic alterations and that these patients also have adverse clinical outcomes (Teschendorff, Gao, et al., 2016). An association between outlier methylation levels and recurrent miscarriage has also been noted (Hanna, McFadden, & Robinson, 2013), as have outlier methylation signatures that are tumor-type-specific (Gaykalova et al., 2015).

All of these studies suggest that some individuals behave as so-called miner's canaries for individual environmental exposures and may be peculiarly susceptible to disruption of the epigenome by particular environmental factors (even when mechanism is not clear). If particular classes of cancer patients can be stratified by different environmental exposures, based on their epigenetic signatures in normal tissues, one might distinguish mechanistic subclasses of the same tumor type. We hypothesized that the potential association of ART and low birth weight children and the corresponding subsequent overweight/obesity among adults who were low birth weight children (Ong & Dunger, 2002; Varvarigou, 2010) might identify a characteristic fraction of breast and colon cancer patients (Bousquenaud, Fico, Solinas, Rüegg, & Santamaria-Martínez, 2018; Laiyemo, 2014; Zeng & Lazarova, 2012). The identification of additional series of associations among cancer patients would be of similar interest.

3. OMP patients in TCGA

We analyzed TCGA datasets for the identification of OMP individuals in nine types of cancer (Fig. 3, Table 1). These nine tumor types were selected because we restricted our analysis to individuals for which data were available from the Illumina Infinium HumanMethylation450 BeadChip (450K array) because of the even distribution of target CpG sites with

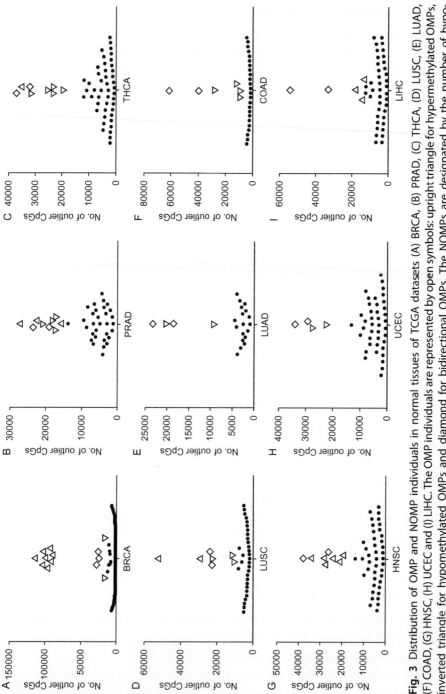

Fig. 3 Distribution of OMP and NOMP individuals in normal tissues of TCGA datasets (A) BRCA, (B) PRAD, (C) THCA, (D) LUSC, (E) LUAD, (F) COAD, (G) HNSC, (H) UCEC and (I) LIHC. The OMP individuals are represented by open symbols: upright triangle for hypermethylated OMPs, inverted triangle for hypomethylated OMPs and diamond for bidirectional OMPs. The NOMPs are designated by the number of hypomethylated outlier CpGs for BRCA, THCA, LUSC, LUAD and COAD; and by the number of hypermethylated outlier CpGs for PRAD, HNSC, UCEC and LIHC. The hypermethylated and hypomethylated OMPs are designated by the number of hypermethylated and hypomethylated outlier CpGs, respectively. Bidirectional OMPs are designated by the greater of the number of hypermethylated or hypomethylated outlier CpGs.

Table 1 Frequency of OMP Individuals in TCGA datasets.

TCGA dataset	Primary tumor tissue	No. samples	OMPs N(%)	Hypermethylated OMPs N(%)	Hypomethylated OMPs N(%)	Bidirectional OMPs N(%)
BRCA	Breast	95	13(13.68)	8(8.42)	2(2.11)	3(3.16)
PRAD	Prostate	40	9(22.50)	3(7.50)	4(10.00)	2(5.00)
THCA	Thyroid	54	8(14.81)	1(1.85)	5(9.26)	2(3.70)
LUSC	Lung	38	7(18.42)	2(5.26)	3(7.89)	2(5.26)
LUAD	Lung	22	4(18.18)	0(0.00)	2(9.09)	2(9.09)
COAD	Colon	30	6(20.00)	0(0.00)	4(13.33)	2(6.67)
HNSC	Head & neck	50	8(16.00)	6(12.00)	0(0.00)	2(4.00)
UCEC	Uterus	45	4(8.89)	0(0.00)	2(4.44)	2(4.44)
LIHC	Liver	45	5(11.11)	3(6.67)	0(0.00)	2(4.44)

OMP refers to Outlier Methylation Phenotype in Normal Tissues taken from the same organ in which the tumor occurred.

low (0–20%), intermediate (20–80%) and high (80–100%) methylation levels. The distribution of low, intermediate and high methylation range CpGs were 37–40%, 26–33% and 28–35%, respectively, across the analyzed TCGA datasets. The frequency of OMP individuals among different cancer patients' tumor types ranged from 8.89% to 22.50% (Table 1). All datasets have both hypermethylated and hypomethylated outlier methylation levels OMP individuals, as well as individuals who are OMP in both directions (Table 1) that is, outliers with abnormally disrupted hypermethylated CpGs (Hypermethylated OMPs) as well as abnormally disrupted hypomethylated CpGs (Hypomethylated OMPs). Most of the datasets have more hypomethylated OMPs compared to hypermethylated OMPs. However, breast, head and neck, and liver cancers have more hypermethylated OMPs. Only a small fraction of OMP individuals were both hypermethylated OMPs and hypomethylated OMPs (Bidirectional OMPs) in all the cancer datasets. This defines three distinct types of OMP individuals based on the direction of outlier methylation: Hypermethylated OMPs, Hypomethylated OMPs and Bidirectional OMPs. The hypermethylated OMPs are mostly outliers/abnormally methylated in the hypomethylated regions of the genome that spans the CpG islands (CGIs) and their nearby domains (shores and shelves). Also, these areas are predominantly the promoter regions of the genes associated with transcriptional activity. On the other hand, the hypomethylated OMPs are outliers/abnormally methylated in hypermethylated regions of the genome, encompassing the gene bodies (such as exons) and affect the gene products. Interestingly, the bidirectional OMPs have disrupted epigenomes in both CGIs and gene bodies. Hence, it is a possibility that bidirectional OMPs might be more prone to extreme transcriptional consequences, compared to the other two counterparts. Interestingly, in a recent analysis of TCGA individuals within a tumor type for whom methylation profiles did not cluster them with their counterparts (Mamatjan et al., 2017), these tumor type methylation outliers were also characterized as outliers based on their transcriptional profiles. Classification of OMP individuals based on the directionality of outlier methylation, as well as presence of specific mutations, might suggest causes and potential consequences of OMP in cancer datasets.

4. Do OMP patients have CIMP tumors?

As noted earlier, CIMP and OMP each define a molecular phenotype in tumor and normal tissues, respectively, and the most obvious question is whether OMP individuals are more likely to develop CIMP positive tumors. We interrogated the overlap of OMPs in normal tissue with CIMP positive tumors using previously published CIMP categorizations (Sánchez-Vega, Gotea, Margolin, & Elnitski, 2015) for the TCGA datasets run on the Illumina 450 K methylation array platform. In this study, unsupervised clustering was used to stratify 12 TCGA cancer datasets into CIMP positive and CIMP negative subpopulations. An intermediate group, CIMP intermediate was also identified in this study. The stratification into different CIMP subtypes was based on differential methylation of the probes in normal and tumor tissues. The selected probes have low methylation in normal tissues (average methylation below 5%) and increased methylation in the tumor tissues (average methylation above 25%). This filtering criteria for probes yielded zero probes (i.e., no CIMP positive tumors) for one of the datasets (thyroid carcinoma—THCA).

The distribution of CIMP positive samples among OMP individuals and NOMP (Non Outlier Methylation Phenotype) individuals was not significantly different in any of the TCGA datasets (Table 2). Breast invasive carcinoma (BRCA) showed the highest overlap between CIMP positive tumors and OMP individuals (30.77%), whereas LUng Squamous Cell carcinoma (LUSC) had no overlap of OMP individuals with CIMP positive tumors (Table 2). Interestingly, all the CIMP positive OMPs were

Table 2 Distribution of CIMP+ in TCGA datasets.

TCGA dataset	OMPs N[a]	NOMPs N[a]	CIMP+ in OMPs N(%)	CIMP+ in NOMPs N(%)	P value[b]
BRCA	13	76	4(30.77)	19(25.00)	0.7342
PRAD	9	31	1(11.11)	5(16.13)	1.000
LUSC	5	31	0(0.00)	6(19.35)	0.5638
LUAD	4	15	1(25.00)	1(6.67)	0.3860
COAD	6	24	1(16.67)	8(33.33)	0.6371
HNSC	8	42	1(12.50)	10(23.81)	0.6658
UCEC	4	28	1(25.00)	10(35.71)	1.000
LIHC	5	40	1(20.00)	12(30.00)	1.000

[a]Number of individuals with CIMP data.
[b]Fisher's exact test.
CIMP+: Individuals positive for CpG Island methylator phenotype; OMPs: individuals with outlier methylation phenotype; NOMPs: individuals with non outlier methylation phenotype.

hypermethylated OMPs in BRCA. Other datasets have only one CIMP positive OMP individual which was hypomethylated OMP in Prostate Adenocarcinoma (PRAD), LUng adenocarcinoma (LUAD) and Colon Adenocarcinoma (COAD); hypermethylated OMP in Head and Neck Squamous Cell carcinoma (HNSC) and; bidirectional OMP in Liver Hepatocellular carcinoma (LIHC) and Uterine Corpus Endometrial carcinoma (UCEC) datasets. These data indicate that CIMP positive tumors do not always develop from OMP normal tissues.

5. Are there pan-cancer OMP CpGs?

We identified OMP individuals in each of the cancer datasets separately. Thus, each dataset has its own set of the most variable CpGs, based on the CpGs that distinguish OMP individuals from NOMP individuals. However, we wished to determine whether such highly variable CpGs (which undergo disruption in OMPs) were specific to each tumor type or whether there exists some common CpGs which undergo abnormal methylation across all datasets, irrespective of the tumor type (pan-cancer OMP CpGs). To address this possibility, we ranked the CpGs based on the number of OMP individuals (hypermethylated, hypomethylated or bidirectional) identified in any tumor type at each site. We used an arbitrary cutoff of 15 OMP individuals (such that >99% CpGs have OMPs from at least four cancer datasets) to designate pan-cancer OMP CpGs. Surprisingly, pan-cancer hypermethylated OMP CpGs (13,918 CpGs) were more than twice as frequent as pan-cancer hypomethylated OMP CpGs (5819 CpGs). It is noteworthy that the total number of hypermethylated OMPs (23) and hypomethylated OMPs (22) was nearly equal (Table 1). This indicates that CpGs with hypermethylated OMPs are more likely to be common across all tissues. As mentioned earlier, these are the CpGs with low endogenous methylation levels that mostly span CGIs and promoter regions. This observation suggests that across all tissues, it is more likely that some of the same CGIs or promoters are more susceptible to epigenetic disruption, resulting in hypermethylated OMPs. Some might argue that the Illumina methylation arrays are over-represented by CGIs or hypomethylated CpGs, so it is expected to see more common CpGs with hypermethylated OMPs. However, as mentioned earlier, we restricted our analyses to 450 K methylation arrays which has a more even distribution of low (37–40% CpGs) and high (28–35% CpGs) methylation level CpGs compared to the 27 K methylation arrays with an overrepresentation of low (>60% CpGs) methylation level CpGs than high (10–15% CpGs) methylation level CpGs.

6. Do OMP patients constitute a distinct subgroup of cancer patients?

We noted earlier that directionality of OMPs might reveal distinct properties of the three OMP (hyper-, hypo- and bidirectional) subtypes. In addition, the magnitude of overall normal tissue genome disruption (what fraction of CpGs had outlier levels of methylation) might also distinguish the OMP individuals. We performed cluster analysis using an R package, Pvclust (Suzuki & Shimodaira, 2006) in each of the cancer datasets to determine whether the OMP individuals cluster together (Fig. 4). Even if OMP individuals cluster together, it is important to note whether they cluster on the basis of direction of abnormal methylation or not. We analyzed three types of clusters in normal tissues: pan-cancer hypermethylated CpGs (13,918 CpGs), pan-cancer hypomethylated CpGs (5819 CpGs) and cancer type-specific, most variable CpGs (~20,000 CpGs) for each dataset. We expected that the clusters based on the pan-cancer hypermethylated and hypomethylated CpGs would determine the importance of OMP directionality in stratifying OMP individuals. In addition, the most variable CpGs based on the number of outliers (irrespective of the direction) for each of the datasets would also give an estimate as to whether the OMPs cluster was based on the direction, magnitude or some unknown phenomenon.

The outcome of the cluster analysis was not uniform for all datasets, but data for three of the cancer types are shown in Fig. 4. Clustering of OMPs in normal tissue samples is shown for breast (Fig. 4A), prostate (Fig. 4C) and thyroid (Fig. 4E). (The question of whether OMP individuals develop tumors with distinct profiles is discussed in the following section with the corresponding tumor tissue data shown in Fig. 4B (breast), D (prostate) and F (thyroid).)

In BRCA, all three cluster analyses resulted in a distinct cluster for all the hypermethylated OMPs (pan-cancer hypermethylated cluster is shown in Fig. 4A). The bidirectional OMPs were also a part of a single cluster in all of the analyses. This observation suggests that hypermethylated and bidirectional OMPs in breast cancer are distinct from hypomethylated OMPs (Fig. 4A).

In PRAD (Fig. 4C), all three cluster analyses had a similar outcome with four of the nine OMP individuals clustering together irrespective of the direction (Fig. 4C). This indicates that in prostate cancer, the magnitude of disrupted methylation overshadows the directionality of the abnormal methylation.

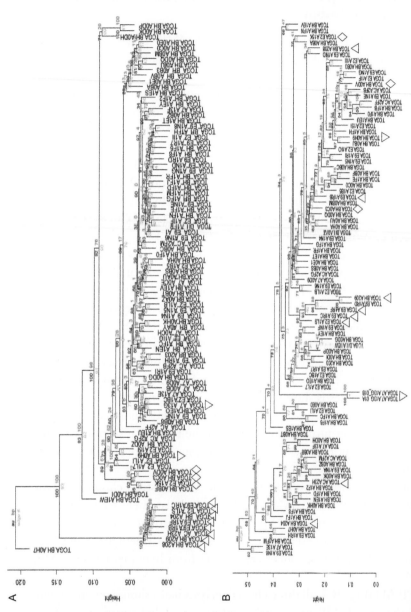

Fig. 4 Cluster dendrograms with AU (approximately unbiased *P* value)/BP (bootstrap probability *P* value) values (%) of methylation patterns in 13,918 pan-hypermethylated CpGs in (A) normal tissues of BRCA, (B) tumor tissues of BRCA, (C) normal tissues of PRAD, (D) tumor tissues of PRAD, (E) normal tissues of THCA and (F) tumor tissues of THCA. The OMP individuals are marked by: upright triangle for hypermethylated OMPs, inverted triangle for hypomethylated OMPs and diamond for bidirectional OMPs.

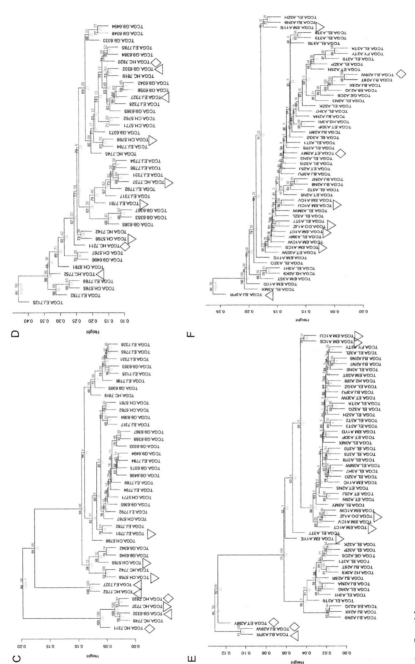

Fig. 4—Cont'd

In THCA (Fig. 4E), the hypermethylated OMP and bidirectional OMPs formed a totally distinct cluster from hypomethylated OMPs and NOMPs in pan-cancer hypermethylated CpGs (Fig. 4E). The hypomethylated OMPs clustered together in the pan-cancer hypomethylated CpGs.

In LUSC, six of the seven OMPs were not a part of any cluster and were totally distinct from the clusters as well as each other in pan-cancer hypermethylated OMPs. This clearly indicates the high variability of the OMPs. Similarly, for LUAD, all the four OMPs appear to be totally distinct from the NOMPs in pan-cancer hypermethylated clustering.

The bidirectional OMPs were totally distinct from the rest of the datasets in COAD, UCEC and LIHC datasets for the pan-cancer CpGs (both hyper- and hypomethylated). The tissue specific CpGs in COAD and LIHC also displayed a similar pattern. In HNSC, few hypermethylated OMPs showed this distinct pattern for the pan-cancer CpGs, whereas for the tissue specific CpGs the OMPs were either distinct or clustered together.

In short, most of the datasets showed distinct clusters for the bidirectional OMPs (BRCA, THCA, COAD, UCEC, LIHC) and hypermethylated OMPs (BRCA, HNSC). The lung datasets (LUSC and LUAD) also showed the variable/distinct nature of the OMPs independent of the direction. These observations suggest that the OMP individuals are clearly distinct from other patients of the same tumor type. Hence, OMP individuals form a distinct class of cancer patients.

7. Do OMP patients have distinct tumor epigenomes?

A significant question is whether the dramatic alterations in DNA methylation at many sites in normal tissue presages any particular genetic alterations in tumors that may arise from the affected field. In other words, do OMPs identified in normal tissues also explain/affect the clustering/methylation patterns in tumor tissues. If normal tissue clustering persists in tumor tissue, then it suggests that the OMPs have normal tissue field effects that may predict the tumor epigenome.

In BRCA, tumors from four of the eight hypermethylated OMPs cluster together for the pan-cancer CpGs (Fig. 4B). In the case of breast cancer specific CpGs, six OMPs were distinct from other samples, suggesting an OMP field-effect in normal breast tissue (Teschendorff, Gao, et al., 2016) that influences additional epigenetic events in tumor tissue.

In PRAD, the OMPs could not be distinguished from NOMPs based on the clustering (Fig. 4D). It is noteworthy that prostate cancer was the most underrepresented in pan-cancer hypermethylated CpGs with only 38.6% CpGs having OMP individuals from this dataset.

In THCA, the single hypermethylated OMP was distinct from rest of the samples for pan-cancer hypermethylated OMPs (Fig. 4F). In pan-cancer hypomethylated CpGs, also one hypermethylated OMP and one bidirectional OMP clustered together to form a totally distinct cluster. Also, three hypomethylated OMPs clustered together.

LUSC, COAD and LIHC datasets indicated high variability of the tumors of OMP individuals as one hypomethylated OMP; one hypomethylated OMP and one bidirectional OMP were distinct from rest of the patients in separate cluster dendrograms.

In HNSC and UCEC, only a few of the hypermethylated OMPs and hypomethylated OMPs, respectively, clustered together in at least one of the cluster dendrograms.

In summary, the distinct phenotype of the OMPs was maintained in tumor tissues in some cancer datasets. On the other hand, tumor tissues undergo additional disrupted methylation irrespective of the OMP status in most of the cancer datasets.

8. Are OMP individuals prone to adverse outcomes?

In our discovery study (Ghosh et al., 2016), we found increased prevalence of OMP individuals in the assisted reproduction population (27.3%) compared to the controls (2.8%). The question of whether OMP individuals are also prone to adverse outcomes has been suggested for breast cancer (Teschendorff, Gao, et al., 2016). Some of the TCGA datasets do show a trend of greater numbers of OMP individuals in the cancer subgroups which are at high risk of developing undesirable outcomes. Among COAD patients, the group of right-sided colon COAD patients (i.e., those with lower survival rate) has nearly three-times more OMP individuals than the left-sided COAD group (27.8% versus 9.1%). In the case of BRCA patients, the triple negative group has a slightly higher frequency of OMP individuals than non-triple negative group (11% versus 9%). However, data was not available for five OMP individuals in BRCA.

9. What are potential causes of OMP?

Formally, OMP could be due to genetic effects, environmental exposures or stochastic processes. Almost all individuals show outlier methylation levels at some CpGs and the occurrence of outlier methylation at small numbers of CpGs could be due to chance. However, the occurrence of abnormally methylated CpGs at a vast number (in some cases >100,000 CpGs of the 450,000 profiled) of CpGs, qualifying an individual as OMP, is highly unlikely to be due to chance, leaving environmental exposures or genetic predisposition as the most likely causes. Our original study (Ghosh et al., 2016) associated an environmental effect (in vitro fertilization) with OMP. Genetic predisposition also seems a likely possibility and could result from germline genetic variation in genes that affect DNA methylation directly or indirectly (Bijron et al., 2012; Yang et al., 2016). In fact, multiple trans-acting loci that affect DNA methylation as a quantitative trait have been demonstrated by mapping site-specific DNA methylation levels onto SNP genotype data (Sun, 2014). In this study, the author identified 16,320 meQTLs in peripheral blood lymphocytes. Among these meQTLs, 68% were trans-acting. In this regard, driver mutations have been associated with DNA methylation changes in multiple cancer types (Chen, Gotea, Margolin, & Elnitski, 2017).

We screened the TCGA datasets for mutations/alterations in driver genes (*TP53, CDH1, PIK3CA* in BRCA; *SPOP, TP53* in PRAD; *BRAF, NRAS, HRAS* in THCA; *STK11, KEAP1, TP53* in LUAD; *BRAF, RNF43, MACF1* in COAD; *NSD1, CASP8, TP53* in HNSC; *PTEN, TP53, CTNNB1* in UCEC; *CTNNB1, TP53* in LIHC) as well as for alterations in genes known to affect DNA methylation (DNMTs), demethylation (TETs) enzymes and other reported gene alterations across different cancer types, such as *CDKN2A* in lung cancers (Tam et al., 2013) and *IDH1* and *IDH2* in leukemia (Figueroa et al., 2010). Interestingly, we observed a higher frequency of *CDKN2A* gene alterations in OMPs (86%) compared to NOMPs (32%) in the LUSC dataset. The mutation frequency in NOMPs is concordant with that reported in TCGA (31%). Similarly, we observed higher frequency of gene alterations for several other genes in OMP individuals: *TET3* in PRAD; *DNMT3A* in LUSC; *STK11, TRDMT1* in LUAD; *DNMT3B* in COAD; *IDH1, CASP8* in HNSC; *CUX1* in UCEC. However, none of the gene alterations reached statistical significance, most likely because of the limited number of OMP individuals in individual data sets. However, our study hints at the role such gene mutations/alterations might play in *trans* and influence the abnormal DNA methylation pattern in OMPs.

Another potentially important factor affecting OMP frequency could be epigenetic age (Hannum et al., 2013; Horvath, 2013). OMP individuals

Table 3 Distribution of OMP individuals according to age group in healthy population.

Age group (in years)	No. individuals	No. OMPs N(%)
19–40	38	0(0.00)
40–50	83	6(7.23)
50–60	140	13(9.29)
60–70	167	21(12.57)
70–80	138	25(18.12)
80–90	80	22(27.50)
90–101	10	3(30.00)

OMPs: Individuals with outlier methylation phenotype.

could have greater epigenetic age than non-OMP individuals and thereby be at greater risk for age-related diseases, including cancer (Falandry, Bonnefoy, Freyer, & Gilson, 2014). We interrogated the publicly available methylation data for 656 healthy individuals (GSE40279) to assess the frequency of OMP among individuals without cancer (Hannum et al., 2013). Interestingly, all OMP individuals in this population were greater than 40 years old and the frequency of OMP increased with age, from 7.23% in 40–50-year-olds to 30.00% in 90–100-year-olds (Table 3). These observations suggest that OMP individuals do have greater epigenetic age and may be at greater risk of age-related diseases.

10. Conclusion and future directions

Moving forward, it will be important to identify a much larger number of OMP individuals to better characterize this group and determine which factors influence this phenotype. Although we are unable to identify distinguishing clinical or genetic features of OMP individuals among cancer patients at this time, the unusual nature of the phenotype and its presence in normal tissues is worthy of additional study.

Conflict of interest

None of the authors report any conflict of interest.

References

An, C., Choi, I. S., Yao, J. C., Worah, S., Xie, K., Mansfield, P. F., et al. (2005). Prognostic significance of CpG island methylator phenotype and microsatellite instability in gastric carcinoma. *Clinical Cancer Research, 11*, 656–663.

Arai, E., Chiku, S., Mori, T., Gotoh, M., Nakagawa, T., Fujimoto, H., et al. (2012). Single-CpG-resolution methylome analysis identifies clinicopathologically aggressive CpG island methylator phenotype clear cell renal cell carcinomas. *Carcinogenesis, 33*, 1487–1493.

Bae, Y. K., Brown, A., Garrett, E., Bornman, D., Fackler, M. J., Sukumar, S., et al. (2004). Hypermethylation in histologically distinct classes of breast cancer. *Clinical Cancer Research, 10*, 5998–6005.

Berman, B. P., Weisenberger, D. J., Aman, J. F., Hinoue, T., Ramjan, Z., Liu, Y., et al. (2011). Regions of focal DNA hypermethylation and long-range hypomethylation in colorectal cancer coincide with nuclear lamina-associated domains. *Nature Genetics, 44*, 40–46.

Bijron, J. G., van der Groep, P., van Dorst, E. B., Seeber, L. M., Sie-Go, D. M., Verheijen, R. H., et al. (2012). Promoter hypermethylation patterns in fallopian tube epithelium of BRCA1 and BRCA2 germ line mutation carriers. *Endocrine-Related Cancer, 19*, 69–81.

Bousquenaud, M., Fico, F., Solinas, G., Rüegg, C., & Santamaria Martínez, A. (2018). Obesity promotes the expansion of metastasis-initiating cells in breast cancer. *Breast Cancer Research, 20*, 104.

Chen, Y. C., Gotea, V., Margolin, G., & Elnitski, L. (2017). Significant associations between driver gene mutations and DNA methylation alterations across many cancer types. *PLoS Computational Biology, 13*, e1005840.

Falandry, C., Bonnefoy, M., Freyer, G., & Gilson, E. (2014). Biology of cancer and aging: A complex association with cellular senescence. *Journal of Clinical Oncology, 32*, 2604–2610.

Fang, F., Turcan, S., Rimner, A., Kaufman, A., Giri, D., Morris, L. G., et al. (2011). Breast cancer methylomes establish an epigenomic foundation for metastasis. *Science Translational Medicine, 3*, 75ra25.

Figueroa, M. E., Abdel-Wahab, O., Lu, C., Ward, P. S., Patel, J., Shih, A., et al. (2010). Leukemic IDH1 and IDH2 mutations result in a hypermethylation phenotype, disrupt TET2 function, and impair hematopoietic differentiation. *Cancer Cell, 18*, 553–567.

Gaykalova, D. A., Vatapalli, R., Wei, Y., Tsai, H. L., Wang, H., Zhang, C., et al. (2015). Outlier analysis defines zinc finger gene family DNA methylation in tumors and saliva of head and neck cancer patients. *PLoS One, 10*, e0142148.

Ghosh, J., Mainigi, M., Coutifaris, C., & Sapienza, C. (2016). Outlier DNA methylation levels as an indicator of environmental exposure and risk of undesirable birth outcome. *Human Molecular Genetics, 25*, 123–129.

Goelz, S. E., Vogelstein, B., Hamilton, S. R., & Feinberg, A. P. (1985). Hypomethylation of DNA from benign and malignant human colon neoplasms. *Science, 228*, 187–190.

Hanna, C. W., McFadden, D. E., & Robinson, W. P. (2013). DNA methylation profiling of placental villi from karyotypically normal miscarriage and recurrent miscarriage. *The American Journal of Pathology, 182*, 2276–2284.

Hannum, G., Guinney, J., Zhao, L., Zhang, L., Hughes, G., Sadda, S., et al. (2013). Genome-wide methylation profiles reveal quantitative views of human aging rates. *Molecular Cell, 49*, 359–367.

Hansen, K. D., Timp, W., Bravo, H. C., Sabunciyan, S., Langmead, B., McDonald, O. G., et al. (2011). Increased methylation variation in epigenetic domains across cancer types. *Nature Genetics, 43*, 768–775.

Hinoue, T., Weisenberger, D. J., Lange, C. P., Shen, H., Byun, H. M., Van Den Berg, D., et al. (2012). Genome-scale analysis of aberrant DNA methylation in colorectal cancer. *Genome Research, 22,* 271–282.

Horvath, S. (2013). DNA methylation age of human tissues and cell types. *Genome Biology, 14,* R115.

Hughes, L. A., Melotte, V., de Schrijver, J., de Maat, M., Smit, V. T., Bovée, J. V., et al. (2013). The CpG island methylator phenotype: What's in a name? *Cancer Research, 73,* 5858–5868.

Laiyemo, A. O. (2014). The risk of colonic adenomas and colonic cancer in obesity. *Best Practice & Research. Clinical Gastroenterology, 28,* 655–663.

Li, X., Qiu, W., Morrow, J., DeMeo, D. L., Weiss, S. T., Fu, Y., et al. (2015). A comparative study of tests for homogeneity of variances with application to DNA methylation data. *PLoS One, 10* e0145295.

Mainigi, M. A., Sapienza, C., Butts, S., & Coutifaris, C. (2016). A molecular perspective on procedures and outcomes with assisted reproductive technologies. *Cold Spring Harbor Perspectives in Medicine, 6,* a023416.

Mamatjan, Y., Agnihotri, S., Goldenberg, A., Tonge, P., Mansouri, S., Zadeh, G., et al. (2017). Molecular signatures for tumor classification: An analysis of the cancer genome atlas data. *The Journal of Molecular Diagnostics, 19,* 881–891.

Maruyama, R., Toyooka, S., Toyooka, K. O., Harada, K., Virmani, A. K., Zöchbauer-Müller, S., et al. (2001). Aberrant promoter methylation profile of bladder cancer and its relationship to clinicopathological features. *Cancer Research, 61,* 8659–8663.

Maruyama, R., Toyooka, S., Toyooka, K. O., Virmani, A. K., Zöchbauer-Müller, S., Farinas, A. J., et al. (2002). Aberrant promoter methylation profile of prostate cancers and its relationship to clinicopathological features. *Clinical Cancer Research, 8,* 514–519.

Miller, B. F., Sánchez-Vega, F., & Elnitski, L. (2016). The emergence of Pan-cancer CIMP and its elusive interpretation. *Biomolecules, 6.*

Noushmehr, H., Weisenberger, D. J., Diefes, K., Phillips, H. S., Pujara, K., Berman, B. P., et al. (2010). Identification of a CpG island methylator phenotype that defines a distinct subgroup of glioma. *Cancer Cell, 17,* 510–522.

Ogino, S., Kawasaki, T., Kirkner, G. J., Kraft, P., Loda, M., & Fuchs, C. S. (2007). Evaluation of markers for CpG island methylator phenotype (CIMP) in colorectal cancer by a large population-based sample. *The Journal of Molecular Diagnostics, 9,* 305–314.

Ong, K. K., & Dunger, D. B. (2002). Perinatal growth failure: The road to obesity, insulin resistance and cardiovascular disease in adults. *Best Practice & Research Clinical Endocrinology & Metabolism, 16,* 191–207.

Sánchez-Vega, F., Gotea, V., Margolin, G., & Elnitski, L. (2015). Pan-cancer stratification of solid human epithelial tumors and cancer cell lines reveals commonalities and tissue-specific features of the CpG island methylator phenotype. *Epigenetics & Chromatin, 8,* 14.

Shen, L., Toyota, M., Kondo, Y., Lin, E., Zhang, L., Guo, Y., et al. (2007). Integrated genetic and epigenetic analysis identifies three different subclasses of colon cancer. *Proceedings of the National Academy of Sciences of the United States of America, 104,* 18654–18659.

Strathdee, G., Appleton, K., Illand, M., Millan, D. W., Sargent, J., Paul, J., et al. (2001). Primary ovarian carcinomas display multiple methylator phenotypes involving known tumor suppressor genes. *The American Journal of Pathology, 158,* 1121–1127.

Sun, Y. V. (2014). The influences of genetic and environmental factors on methylome-wide association studies for human diseases. *Current Genetic Medicine Reports, 2,* 261–270.

Sunami, E., de Maat, M., Vu, A., Turner, R. R., & Hoon, D. S. (2011). LINE-1 hypomethylation during primary colon cancer progression. *PLoS One, 6* e18884.

Suzuki, M., Shigematsu, H., Iizasa, T., Hiroshima, K., Nakatani, Y., Minna, J. D., et al. (2006). Exclusive mutation in epidermal growth factor receptor gene, HER-2, and KRAS, and synchronous methylation of nonsmall cell lung cancer. *Cancer, 106*, 2200–2207.

Suzuki, R., & Shimodaira, H. (2006). Pvclust: An R package for assessing the uncertainty in hierarchial clustering. *Bioinformatics, 22*, 1540–1542.

Tam, K. W., Zhang, W., Soh, J., Stastny, V., Chen, M., Sun, H., et al. (2013). CDKN2A/p16 inactivation mechanisms and their relationship to smoke exposure and molecular features in non-small-cell lung cancer. *Journal of Thoracic Oncology, 8*, 1378–1388.

Tanemura, A., Terando, A. M., Sim, M. S., van Hoesel, A. Q., de Maat, M. F., Morton, D. L., et al. (2009). CpG island methylator phenotype predicts progression of malignant melanoma. *Clinical Cancer Research, 15*, 1801–1807.

Teschendorff, A. E., Gao, Y., Jones, A., Ruebner, M., Beckmann, M. W., Wachter, D. L., et al. (2016). DNA methylation outliers in normal breast tissue identify field defects that are enriched in cancer. *Nature Communications, 7*, 10478.

Teschendorff, A. E., Jones, A., Fiegl, H., Sargent, A., Zhuang, J. J., Kitchener, H. C., et al. (2012). Epigenetic variability in cells of normal cytology is associated with the risk of future morphological transformation. *Genome Medicine, 4*, 24.

Teschendorff, A. E., Jones, A., & Widschwendter, M. (2016). Stochastic epigenetic outliers can define field defects in cancer. *BMC Bioinformatics, 17*, 178.

Toyota, M., Ahuja, N., Ohe-Toyota, M., Herman, J. G., Baylin, S. B., & Issa, J. P. (1999). CpG island methylator phenotype in colorectal cancer. *Proceedings of the National Academy of Sciences of the United States of America, 96*, 8681–8686.

Toyota, M., Kopecky, K. J., Toyota, M. O., Jair, K. W., Willman, C. L., & Issa, J. P. (2001). Methylation profiling in acute myeloid leukemia. *Blood, 97*, 2823–2829.

Ueki, T., Toyota, M., Sohn, T., Yeo, C. J., Issa, J. P., Hruban, R. H., et al. (2000). Hypermethylation of multiple genes in pancreatic adenocarcinoma. *Cancer Research, 60*, 1835–1839.

Varvarigou, A. A. (2010). Intrauterine growth restriction as a potential risk factor for disease onset in adulthood. *Journal of Pediatric Endocrinology & Metabolism, 23*, 215–224.

Weisenberger, D. J., Siegmund, K. D., Campan, M., Young, J., Long, T. I., Faasse, M. A., et al. (2006). CpG island methylator phenotype underlies sporadic microsatellite instability and is tightly associated with BRAF mutation in colorectal cancer. *Nature Genetics, 38*, 787–793.

Whitcomb, B. P., Mutch, D. G., Herzog, T. J., Rader, J. S., Gibb, R. K., & Goodfellow, P. J. (2003). Frequent HOXA11 and THBS2 promoter methylation, and a methylator phenotype in endometrial adenocarcinoma. *Clinical Cancer Research, 9*, 2277–2287.

Yang, F., Gong, Q., Shi, W., Zou, Y., Shi, J., Wei, F., et al. (2016). Aberrant DNA methylation of acute myeloid leukemia and colorectal cancer in a Chinese pedigree with a MLL3 germline mutation. *Tumour Biology, 37*, 12609–12618.

Zeng, H., & Lazarova, D. L. (2012). Obesity-related colon cancer: Dietary factors and their mechanisms of anticancer action. *Clinical and Experimental Pharmacology & Physiology, 39*, 161–167.

Zhang, C., Li, Z., Cheng, Y., Jia, F., Li, R., Wu, M., et al. (2007). CpG island methylator phenotype association with elevated serum alpha-fetoprotein level in hepatocellular carcinoma. *Clinical Cancer Research, 13*, 944–952.

CHAPTER TWO

Bittersweet tumor development and progression: Emerging roles of epithelial plasticity glycosylations

Ryan M. Phillips[a,†], Christine Lam[a,b,†], Hailun Wang[a,*], Phuoc T. Tran[a,b,c,d,*]

[a]Department of Radiation Oncology and Molecular Radiation Sciences, Sidney Kimmel Comprehensive Cancer Center, Johns Hopkins University School of Medicine, Baltimore, MD, United States
[b]Program in Cellular and Molecular Medicine, Johns Hopkins University School of Medicine, Baltimore, MD, United States
[c]Department of Urology, James Buchanan Brady Urological Institute, Johns Hopkins University School of Medicine, Baltimore, MD, United States
[d]Department of Oncology, Sidney Kimmel Comprehensive Cancer Center, Johns Hopkins University School of Medicine, Baltimore, MD, United States
*Corresponding authors: e-mail address: hwang110@jhmi.edu; tranp@jhmi.edu

Contents

1. Epithelial plasticity—Basics and implications in cancer development, progression and treatment resistance — 24
2. Cancer metabolism—Metabolic adaptations in cancer — 27
3. Glycobiology—An introduction to protein glycosylation and relevance to cancer — 30
4. The hexosamine biosynthesis pathway (HBP)—An emerging metabolic player in cancer — 32
5. Epithelial plasticity and glycosylations in cancer — 35
   5.1 Epithelial plasticity and N-linked glycosylations in cancer — 35
   5.2 Epithelial plasticity and O-linked glycosylations in cancer — 43
6. Associations between EMT and O-GlcNAcylation — 45
7. The EMT-HBP-O-GlcNAcylation axis—An important new pathway to promote the neoplastic phenotype — 48
8. Conclusion and future directions — 54
Acknowledgments — 55
Conflict of interest — 55
References — 55

[†] These authors contributed equally to this work.

Abstract

Altered metabolism is one of the hallmarks of cancer. The best-known cancer metabolic anomaly is an increase in aerobic glycolysis, which generates ATP and other basic building blocks, such as nucleotides, lipids, and proteins to support tumor cell growth and survival. Epithelial plasticity (EP) programs such as the epithelial-mesenchymal transition (EMT) and mesenchymal-epithelial transition (MET) are evolutionarily conserved processes that are essential for embryonic development. EP also plays an important role during tumor progression toward metastasis and treatment resistance, and new roles in the acceleration of tumorigenesis have been found. Recent evidence has linked EMT-related transcriptomic alterations with metabolic reprogramming in cancer cells, which include increased aerobic glycolysis. More recent studies have revealed a novel connection between EMT and altered glycosylation in tumor cells, in which EMT drives an increase in glucose uptake and flux into the hexosamine biosynthetic pathway (HBP). The HBP is a side-branch pathway from glycolysis which generates the end product uridine-5′-diphosphate-N-acetylglucosamine (UDP-GlcNAc). A key downstream utilization of UDP-GlcNAc is for the post-translational modification O-GlcNAcylation which involves the attachment of the GlcNAc moiety to Ser/Thr/Asn residues of proteins. Global changes in protein O-GlcNAcylation are emerging as a general characteristic of cancer cells. In our recent study, we demonstrated that the EMT-HBP-O-GlcNAcylation axis drives the O-GlcNAcylation of key proteins such as c-Myc, which previous studies have shown to suppress oncogene-induced senescence (OIS) and contribute to accelerated tumorigenesis. Here, we review the HBP and O-GlcNAcylation and their putative roles in driving EMT-related cancer processes with examples to illuminate potential new therapeutic targets for cancer.

1. Epithelial plasticity—Basics and implications in cancer development, progression and treatment resistance

Epithelial plasticity (EP) programs such as the epithelial–mesenchymal transition (EMT) and the mesenchymal–epithelial transition (MET) are developmental programs that can engender cells with a fluid cellular phenotype that span fully epithelial to fully mesenchymal properties and potentially many hybrid states in-between (Fig. 1). EP programs are well-conserved, essential for normal development and can be reactivated by some cancer cells (Chaffer & Weinberg, 2011). The EMT has been studied in greater detail and during this process epithelial cells undergo a transcriptional and resultant biochemical change that confers a more mesenchymal phenotype. Extracellular signaling mediated by TGF-β, receptor tyrosine kinases (RTKs),

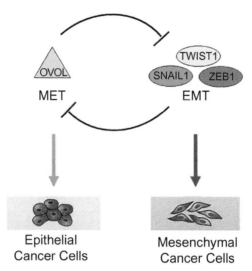

Fig. 1 Model of epithelial plasticity involving EMT-MET balance in human cancer. In human cancer cells the mesenchymal and epithelial states are induced and maintained by transcriptional and post-transcriptional regulatory programs. These programs are controlled by the feedback regulation between the OVOL and EMT-inducing TFs such as Zeb1, Twist1 and Snail1, critical inducers of MET and EMT respectively. Therefore high OVOL and low EMT-TFs stabilize the epithelial state decreasing cancer cell invasion and metastasis, and vice versa for the mesenchymal state.

integrins, WNT, NOTCH, Hedgehog (HH), hypoxia inducible factor 1α (HIF1α) and JAK/STAT induce the EMT in various biological contexts (Taparra, Tran, & Zachara, 2016). Inside cells, canonical EMT is mainly executed by three major families of EMT-inducing transcription factors (TFs): the zinc finger protein Snail family (SNAI1, SNAI2, and SNAI3); the zinc finger E-box binding (ZEB) family; and basic Helix-Loop-Helix (bHLH) proteins (TWIST1 and TWIST2) (Nieto, Huang, Jackson, & Thiery, 2016). These EMT-TFs in general lead to the downregulation of epithelial genes and upregulation of mesenchymal genes, resulting in the loss of cellular adhesion, loss of cell polarity, and reorganization of the cytoskeleton (Carvalho-Cruz, Alisson-Silva, Todeschini, & Dias, 2018). Some canonical markers for this transition are the loss of the epithelial adhesion protein E-cadherin, ZO-1, and cytokeratins and replacement by N-Cadherin, fibronectin, and vimentin. Additional defining cellular phenotypes include a more migratory, stress/death-resistant behavior, immune evasive and neoplastic-prone state of these cells (Dongre et al., 2017; Nieto et al., 2016).

EMT is only half the story of EP as MET-inducing TFs such as OVOL1/2 have been shown to be important regulators of MET in normal and cancer cells (Li & Yang, 2014; Roca et al., 2013). Migratory developmental and cancerous cells both can escape tissues and disseminate to distant organs by EMT, delaminating from the primary organ or tumor, entering the circulation, and transiting to a secondary site. EMT is thought to be a transient program in both contexts, where changes in extrinsic signaling factors induce mesenchymal cells to revert to the epithelial state. MET has been associated with the ability of cancer cells to adapt to a new site, proliferate, and form macrometastases. OVOL TFs orchestrate a transcriptional and splicing-regulatory program that mediates MET, resulting in decreased cancer cell invasiveness by regulating EMT-TFs such as ZEB1. This "EP hypothesis" is supported by the observation that metastases originating from epithelial cancers show a high degree of differentiation toward the epithelial phenotype suggesting epithelial metastatic cells undergo an EMT-to-MET to facilitate macroscopic metastatic colonization, although much work is required to validate and further substantiate this hypothesis (Jia et al., 2015; Tsai, Donaher, Murphy, Chau, & Yang, 2012).

EP programs such as EMT have been established to be important late in the neoplastic process for metastasis depending on the cancer type and other contextual cues (Brabletz, Brabletz, & Stemmler, 2017; Chen et al., 2018; Krebs et al., 2017; Zheng et al., 2015). However, an equally important activity of these EP programs directed by EMT-inducing TFs occurs very early during epithelial tumorigenesis for the suppression of tumor suppressor programs such as oncogene-induced senescence (OIS) and oncogene-induced apoptosis (OIA) (Puisieux, Brabletz, & Caramel, 2014; Tran et al., 2012). TWIST1 and SNAIL1 have both independently been demonstrated to accelerate tumorigenesis by inhibiting Rb and p53 tumor suppressor pathways (Ansieau et al., 2008; Lee & Bar-Sagi, 2010; Ni et al., 2016). Both TWIST1 and SNAIL1 have been shown to antagonize p53 directly (Ni et al., 2016; Shiota et al., 2008). Snail1 has been shown to be in a complex with wildtype, but not mutant p53, resulting in p53 deacetylation and increased p53 degradation. Using a combination of Snail1- and p53-deficient autochthonous breast tumor and in vitro derived organoid models, Snail1has been established as a critical repressor of p53-mediated tumor suppression in a luminal B subtype breast cancer mouse model. There are also p53-independent pathways for senescence suppression by TWIST1 in lung cancer cells which we have documented (Burns et al., 2013; Yochum et al., 2017). Alain Puisieux's group has recently further added to this paradigm of

EP factors contributing to early tumorigenesis in breast cancer (Morel et al., 2017). Careful cell sorting and genetic manipulation ex vivo of breast epithelial cell populations from reduction mammoplasties suggested only specific breast stem-cell populations tolerate oncogenic activation and thus have increased susceptibility to neoplastic transformation. All together, these data suggest the importance of cellular reprogramming by specific EMT-inducing TFs in defined cellular lineages for particular oncogenic insults during early tumorigenesis (Puisieux, Pommier, Morel, & Lavial, 2018).

Importantly, it has been well documented that cancer cells surviving treatment are enriched in EMT markers, and several recent studies presented data that directly link the EMT to cancer drug resistance, which may be caused by an enhancement of cancer cell survival, cell fate transition, cancer stem cells generation, and/or up-regulation of drug resistance-related genes (Arumugam et al., 2009; Fischer et al., 2015; Mani et al., 2008; Shintani et al., 2011; Thomson, Petti, Sujka-Kwok, Epstein, & Haley, 2008; Yochum et al., 2019; Zheng et al., 2015). In summary, EP programs of EMT and MET appear to play major roles in the pathology and clinical behavior of many epithelial cancers and thus a heightened molecular understanding of these programs may afford promising strategies to treat the full spectrum of the neoplastic phenotype (Malek, Wang, Taparra, & Tran, 2017).

2. Cancer metabolism—Metabolic adaptations in cancer

Early investigation into tumor metabolism was based on the observation that cancer cells grow very rapidly and in a disorganized manner, thereby setting them apart from either the rapidly proliferative but highly organized cells of normal development or the relatively less active normal epithelia (Warburg, 1925). Warburg and colleagues observed that, contrary to their expectation, carcinoma tissue exhibited levels of respiration lower than normal kidney or liver tissue, but underwent anaerobic glycolysis, the transformation of simple sugars to lactic acid in the absence of oxygen, at a rate many-fold higher than normal tissues. Subsequently, they observed that carcinoma tissue maintains significant rates of glycolysis even under aerobic conditions, while normal tissues do not. In a follow-up publication, Warburg et al. showed that carcinoma cells could survive not only under anaerobic conditions permitting glycolysis, but not respiration, and also when deprived of glucose under aerobic conditions, but not in circumstances precluding both glycolysis and respiration concurrently (Warburg,

Wind, & Negelein, 1927). These findings set the stage for investigation into whether the upregulation of glycolysis is merely to ensure energy production during oxygen stress, or if this metabolic derangement provides additional selective advantages to cancer cells.

The initial result of glycolysis is the conversion of a single molecule of glucose into 2 of pyruvate, a net increase of 2 ATP, and the reduction of 2 NAD+ to NADH. The subsequent fate of pyruvate varies based on environmental factors, most importantly in the presence of oxygen: pyruvate can be converted anaerobically into lactate ("anaerobic glycolysis") or aerobically into acetyl-CoA which can then proceed through the citric acid cycle and fuel mitochondrial oxidative phosphorylation. Energetically, anaerobic glycolysis results in a redox-neutral production of 2 ATP, while progression through oxidative phosphorylation ultimately produces 30–32 ATP per glucose consumed; it is this latter process that typically predominates in normal human tissues. The "aerobic glycolysis" described by Warburg and colleagues refers not to a unique process, but rather a tendency of cancer cells (or other rapidly proliferative cells) to convert large amounts of pyruvate to lactate anaerobically even in the presence of oxygen.

Decades of debate and investigation were spent asking whether this shift toward glycolysis was because tumors cannot appropriately utilize oxidative phosphorylation due to mitochondrial dysfunction, or if the biosynthetic benefits of glycolysis balance out the significant energetic sacrifices incurred. Ultimately the latter assertion appears most consistent with our collective understanding of tumor biology. While fewer ATP molecules are generated per glucose than through oxidative phosphorylation, the conversion of pyruvate to lactate can lead to faster ATP production and NAD+ regeneration than oxidative phosphorylation (Guppy, Greiner, & Brand, 1993). Maintenance of a sufficient intracellular glucose pool to fuel glycolysis is achieved by significant upregulation of glucose transporters by tumor cells (Labak et al., 2016), a phenomenon which is evident in the broad diagnostic utility of positron emission tomography (PET) imaging using ^{18}F-fluorodeoxyglucose (^{18}F-FDG) in cancer detection. Additionally, tumor upregulation of enzymes involved in gluconeogenesis, the production of glucose from smaller organic molecules which can be broadly conceptualized as "reverse glycolysis," provides an important mechanism for maintaining glucose pools. In principle, "reverse glycolysis" may be particularly relevant to areas in the body which are poorly perfused and thus less able to scavenge glucose from the bloodstream (Leithner et al., 2015). Furthermore, a tumor requires not just energy, but all of the building blocks necessary for growth (Vander Heiden, Cantley, & Thompson, 2009).

Far beyond glycolysis, there is a large body of literature describing how oncogenes and tumor suppressor genes commonly implicated in tumorigenesis modify the bioenergetics of cancer cells to promote anabolism necessary for rapid growth proliferation (Jones & Thompson, 2009). Upregulation of c-Myc leads to increased aerobic glycolysis by increasing lactate dehydrogenase (LDH-A) activity (Osthus et al., 2000), as well as increased nucleotide and amino acid metabolism (Gordan, Thompson, & Simon, 2007). The pentose phosphate pathway, also known as the phosphogluconate pathway (or hexose monophosphate shunt), metabolizes glucose-6-phosphate to ribulose-5-phosphate, a precursor of ribonucleotide synthesis, while generating reduced nicotinamide adenine dinucleotide phosphate (NADPH), critical for maintenance of cellular redox balance and a component of several anabolic pathways (Kowalik, Columbano, & Perra, 2017). These products are vitally important to highly proliferative cells and several of the enzymes in this pathway including glucose-6-phosphate dehydrogenase (G6PD), 6-phosphogluconate dehydrogenase, transaldolase, and transketolase can be overactive in cancers. Tumor cells are also known to upregulate de novo synthesis of fatty acids, mobilization of fatty acid stores from lipolysis of triglycerides, and uptake of exogenous lipids from their surroundings through increased expression of the fatty acid receptor CD36, low density lipoprotein receptor (LDLr), and the fatty acid binding proteins FABP4, FABP5, and FABP7 (Liu, Luo, Halim, & Song, 2017).

More recent evidence links EMT-related transcriptomic alterations with metabolic reprograming in cancer cells, which include increased aerobic glycolysis and the accumulation of dihydropyrimidines (Shaul et al., 2014). EMT has been implicated in metabolic reprogramming of cancer cells; cancer cells undergoing EMT exhibit aberrant glucose metabolism, but these alterations have yet to be vigorously explored (Morandi, Taddei, Chiarugi, & Giannoni, 2017). It has been shown previously that cancer cells undergoing EMT increase glucose intake; Liu et al. exposed pancreatic ductal carcinoma (PDAC) cells to the EMT inducers TNFα and TGF-β and found increased glucose intake and lactate production with no effect on oxidative phosphorylation (OXPHOS) metabolism (Liu, Quek, Sultani, & Turner, 2016). Dong et al. demonstrated that EMT in breast cancer exhibited glycolysis induction, increased glucose uptake, and inhibition of oxygen consumption (Dong et al., 2013); Kondaveeti et al. also induced EMT in two breast cancer cell lines and found increased glycolysis and decreased gluconeogenesis (Kondaveeti, Guttilla Reed, & White, 2015). Two recent studies also revealed a novel connection between EMT and altered glycosylation in tumor cells. Lucena et al. demonstrated an increase in glucose uptake and flux into the hexosamine biosynthetic pathway (HBP) during TGF-β-induced EMT in a non-small cell lung cancer (NSCLC) cell line (Lucena et al., 2016). In addition, Taparra et al. found that EMT elevated levels of key HBP genes in NSCLC mouse models (Taparra et al., 2018). These studies highlight that glucose metabolism, especially the HBP, may be important metabolic pathways to target in EMT-mediated cancers and will be discussed in greater detail below.

3. Glycobiology—An introduction to protein glycosylation and relevance to cancer

Glycosylation is the covalent attachment of oligo- or polysaccharide sugars to other molecules such as proteins to form glycans and glycoproteins. Glycosylation of other macromolecules such as lipids and even other sugars occurs abundantly in nature, but our focus will be on protein glycoconjugates. Glycosylation is one of the most important biological post-translational protein modifications and occurs predominantly in the cytosol, the endoplasmic reticulum, the Golgi apparatus and at the cell membrane. Glycosylation of proteins can substantially influence and modulate protein structure and function and appears to be involved in the fine tuning of cell–cell recognition and cellular signaling processes. Interestingly, altered glycosylation is a universal feature of cancer cells, and has been frequently correlated with malignant transformation and tumor progression. Glycosylations are well positioned to mediate many of the cellular changes associated with cancer, modifying cell–cell adhesion, responsiveness to growth factors, immune system evasion, metabolic adaptation and changes in signal transduction (Varki, Kannagi, Toole, & Stanley, 2015).

In general, there are two major types of glycosylation that occur on proteins. More commonly carbohydrate moieties can be linked to the amide group of asparagine (N-linkage) or the hydroxyl-containing sidechain of serine or threonine (O-linkage). N-linked glycosylation is a frequent processing step for proteins bound for the plasma membrane or export from the cell. The initial step takes place within the lumen of the endoplasmic reticulum, where NXS/T motifs (N: asparagine, X: any amino acid but proline, S/T: serine or threonine) are N-glycosylated by oligosaccharyltransferase (OST), which transfers a highly conserved Glc_3Man_9 $GlcNAc_2$ oligosaccharide to the asparagine sidechain of the acceptor motif (Aebi, 2013). After assembly and conjugation of the core oligosaccharide in the ER, a wide variety of additional modifications can occur in the Golgi apparatus, including substitution and/or addition of saccharides moieties as well as creation of higher-order branching structures (Wang et al., 2017). The resulting N-linked glycoproteins are classified as high mannose, complex, and hybrid. The core structure of all N-linked glycoproteins is a chain of two N-acetylglucosamine (GlcNAc) followed by a branching mannose triad. High mannose N-linked glycoproteins are typified by additional branching mannose residues beyond the core structure, while complex N-linked glycoproteins may contain additional carbohydrate moieties including but certainly not limited to glucose, galactose, fucose, xylose, GlcNAc, N-acetylgalactosamine, N-acetylneuraminic acid (commonly known as sialic acid), N-glycolylneuraminic acid, and glucuronic acid (Costa, Rodrigues, Henriques, Oliveira, & Azeredo, 2014). Relevant to this review, many prototypical EP markers are modified by N-linked glycans such as E-cadherin and N-cadherin which ultimately affect their protein activity and function resulting in changes in cellular signal transduction, adhesion and migration (Guo, Lee, Kamar, & Pierce, 2003; Pinho et al., 2009; Zhou et al., 2008). In addition, N-glycosylation of cell surface receptors is an important determinant of signaling behavior as the number and structure of N-glycans on receptors influences their rate of endocytosis (Dennis, Lau, Demetriou, & Nabi, 2009). For example, association of N-glycans on unbound receptor tyrosine kinases with galectins in the plasma membrane restricts lateral diffusion, prevents endocytosis, and thereby promotes availability for ligand binding, but also serves to prevent ligand-independent receptor activation and maintain signaling integrity. This concept extends beyond the cell of origin, as glycosylation also impacts protein interactions with distant cells and, in higher organisms such as humans, removal of target cells from the circulation (Costa et al., 2014).

O-linked glycosylation is more variable in its execution than N-linked glycosylation, but no less important. Among the simplest forms of O-linked glycosylation is the addition of a single mannose to the hydroxyl of a serine or threonine residue, termed O-mannosylation, appears to be involved in identification and processing of misfolded proteins within the endoplasmic reticulum (Loibl & Strahl, 2013). Conjugation of a single GlcNAc, referred to as O-GlcNAcylation, acts as a rapidly cycling modification more conceptually akin to phosphorylation than to complex glycosylation and regulates transcription, translation, and protein trafficking within the nucleus, cytoplasm and mitochondria (Hart & Akimoto, 2009). For example, increased O-GlcNAcylation of G6PD in non-small cell lung cancer promotes growth and proliferation through upregulation of the pentose phosphate pathway activity (Rao et al., 2015). O-GlcNAcylation is a unique form of protein glycosylation that will be discussed in greater detail below. More complex, branching O-glycans are frequently antigenic, for example the determinants of the ABO and Lewis blood groups, and are integral to the mucinous glycoproteins found in mucous membrane secretions (Brockhausen, Schachter, & Stanley, 2009).

4. The hexosamine biosynthesis pathway (HBP)—An emerging metabolic player in cancer

The HBP is a side-branch metabolic sensing pathway from glycolysis, in which 2–5% of total glucose is diverted at the fructose-6-phosphate step. The HBP ultimately converts glucose to the end product uridine-5′-diphosphate-N-acetylglucosamine (UDP-GlcNAc), which is critical for the post-translational modifications of proteins, such as protein glycosylation. Despite the limited flux through the HBP, cellular UDP-GlcNAc is among the most abundant high-energy cellular compounds reaching levels over 1 mM in concentration (Sasai, Ikeda, Fujii, Tsuda, & Taniguchi, 2002). The first step in the HBP is the transfer of an amide group by glutamine:fructose-6-phosphate amidotransferase (GFAT) from glutamine to fructose-6-phosphate to form glucosamine-6-phosphate (GlcN6P) and glutamate (Fig. 2A) (Marshall, Bacote, & Traxinger, 1991). This step is rate-limiting for the HBP and the rapid turnover ($t_{1/2} = 45$ min) allows for rapid adaptation to environmental alterations; furthermore, the action of GFAT is the only known mechanism for production of GlcN-6P. Feedback inhibition of GFAT activity occurs through allosteric binding of UDP-GlcNAc, the terminal product of the HBP, while phosphorylation by protein kinase A (PKA), AMP-activated protein kinase (AMPK) and calcium/calmodulin-dependent kinase II (CaMKII) are implicated in upregulation of GFAT activity, thereby potentially linking the HBP to myriad autocrine, paracrine, and endocrine signaling axes including but not limited to cyclic AMP, adrenergic, adiponectin, and receptor tyrosine kinase signaling (Durand, Golinelli-Pimpaneau, Mouilleron, Badet, & Badet-Denisot, 2008). GFAT activity is also implicated in cytokine synthesis and extracellular matrix interactions (Weigert, Friess, Brodbeck, Haring, & Schleicher, 2003). Conflicting data exist on the role of GFAT isoenzymes in cancer. GFAT1 is a poor prognostic factor in human gastric carcinoma associated with increased invasive behavior and metastasis (Duan et al., 2016). Analysis of fresh gastric carcinoma and normal gastric mucosal tissue identified decreased GFAT1 mRNA and protein expression with concurrent

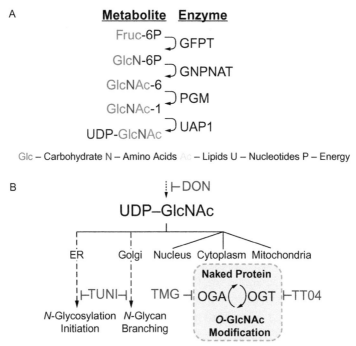

Fig. 2 The hexosamine biosynthetic pathway. (A) Illustration of the metabolites and enzymatic steps of the HBP pathway. Enzymes: GFAT—glutamine:fructose-6-phosphate amidotransferase; GNPNAT—glucosamine-phosphate N-acetyltransferase; PGM—phosphoacetylglucosamine mutase; and UAP1—UDP-N-acetylhexosamine pyrophosphorylase. Metabolites: Fruc-6P—fructose-6-phosphate; GlcN-6P—glucosamine-6-phosphate; GlcNac-6—N-acetylglucosamine-6-phosphate; GlcNac-1—N-acetylglucosamine-1-phosphate; and UDP-GlcNac—UDP-N-acetyl-glucosamine. (B) Schematic showing UDP-GlcNAc incorporation into N-linked and O-linked glycosylation with inhibitors (red).

decreased GlcNAcylation in a subset of tumors, an observation mirrored in representative cell lines and correlated with clinicopathologic variables for poor prognosis such as vascular invasion, advanced T stage, and both nodal and distal metastasis. Kaplan–Meier analysis showed that in both a single-institution patient cohort and data from The Cancer Genome Atlas (TCGA), low GFAT1 expression correlated with inferior overall survival regardless of clinical stage. Knockdown of GFAT1 or chemical inhibition of GFAT1 with 6-diazo-5-oxo-1-norleucine in gastric cancer cells increased invasive behavior and resistance to anoikis, downregulated E-cadherin, increased N-cadherin, vimentin, and Snail1 expression, and increased TGF-β1 secretion. Overall, these data point to GFAT1 as a suppressor of EMT in gastric cancer. Alternatively, *GFPT2* overexpression has been shown to be of prognostic significance in three separate large patient lung adenocarcinoma cohorts for survival and was associated with metabolic reprogramming of the tumor microenvironment and EMT (Zhang et al., 2018).

Following formation of GlcN6P by GFAT, glucosamine-phosphate N-acetyltransferase (GNPNAT) transfers an acetyl group from acetyl-CoA to GlcN-6P to form N-acetylglucosamine-6-phosphate (GlcNAc-6P) and coenzyme A. An alternative to this step involves lysosomal degradation of glycoproteins to form GlcNAc, which can then be phosphorylated by N-acetylglucosamine kinase to form GlcNAc-6P (Boehmelt et al., 2000). Loss of activity in the murine ortholog of GNPNAT has been shown to result in reduced proliferation, cell adhesion, and actin depolymerization, as well as resistance to apoptosis (Boehmelt et al., 2000). These affects appear to be related to defective glycosylation rather than direct action of GNPNAT.

Phosphoacetylglucosamine mutase (PGM3) next transfers the phosphate group of GlcNAac-6P from position 6 to position 1 to form N-acetylglucosamine-1-phosphate (GlcNAc-1P), requiring Mg^{2+} as a cofactor. PGM3 expression is upregulated in human microvascular endothelial cells in response to erythropoietin, suggesting a role in angiogenesis, again likely through modulation of interactions between cells and the extracellular matrix (Li, Rodriguez, & Banerjee, 2000). Congenital mutations resulting in PGM3 loss-of-function lead to glycosylation defects manifesting as decreased CD8+ T-cells and increased production of the T_H2 cytokines IL-4, IL-5, IL-13, and IL-17 by stimulated $CD4^+$ T-cells (Zhang et al., 2014). A recently described synthetic inhibitor of PGM3 was found to cause significant decreases in tumor proliferation and survival in breast cancer cell lines and mouse xenograft models (Ricciardiello et al., 2018).

The final step of the HBP is the conjugation of a uridine diphosphate (UDP) moiety to GlcNAc-1P by UDP-N-acetylhexosamine pyrophosphorylase (UAP1) to form UDP-N-acetyl-glucosamine (UDP-GlcNAc). This process requires uridine triphosphate (UTP) as a reactant, Mg^{2+}/Mn^{2+} as a cofactor, and produces pyrophosphate as a byproduct (Peneff et al., 2001). UAP1 is also capable of producing UDP-N-acetyl-galactosamine-1-phosphate (UDP-GalNAc) from N-acetyl-galactosamine-1-phosphate (GalNAc-1P), with the relative affinities of UAP1 for the glucosamine or galactosamine-based species dictated by alternative splicing (Wang-Gillam, Pastuszak, & Elbein, 1998). UAP1 overexpression has recently been identified in prostate cancer cell lines and appears to protect cells against the growth inhibitory effects of N-linked glycosylation inhibitors (Itkonen et al., 2015).

UDP-GlcNac is one of the essential carbohydrate building blocks of the $Glc_3Man_9GlcNAc_2$ oligosaccharide transferred during the ER phase of glycosylation (Aebi, 2013). Subsequent modification of N-glycans in the Golgi apparatus is also dependent on the presence of UDP-GlcNAc and HBP dynamics are tightly linked to cell growth, differentiation, and surface receptor presentation (Lau et al., 2007). Interestingly, the HBP is well positioned to serve as a biochemical metabolic barometer for the cell sensing the four macromolecules of life, coordinating carbohydrate, amino acid, lipid, and nucleotide donors through incorporation of Fru-6P, Gln, acetyl-CoA, and uridine, respectively (Wells, Vosseller, & Hart, 2003).

Arguably the most critical downstream utilization of UDP-GlcNAc is the nutrient- and stress-responsive posttranslational modification, O-GlcNAcylation, which involves the attachment of a single GlcNAc moiety to Ser and Thr residues of over 3000 intra-cellular proteins via a β-glycosidic bond (Wells et al., 2003). In contrast to N-linked glycans, which are controlled by upward of almost two dozen separate enzymes and can be highly branched structures, a single pair of enzymes—O-GlcNAc transferase (OGT) and O-GlcNAcase (OGA)—controls the addition and removal of a single O-GlcNAc (Iyer & Hart, 2003). O-GlcNAcylation competes with phosphorylation on Ser/Thr and thus regulates diverse cellular processes, which include transcription, protein stability and cell signaling dynamics. Recent studies also indicated that global changes in protein O-GlcNAcylation is emerging as a general characteristic of cancer cells.

5. Epithelial plasticity and glycosylations in cancer

A growing body of evidence links abnormal glycosylation with EP in cancer cells, and suggest that cells engaged in EMT-MET undergo a metabolic reprograming process which fuels aberrant glycosylation patterns in cancer cells. Here we comprehensively review the most germane examples from the literature putting them into the context of tumor development and progression toward metastasis beginning with N-linked glycans and concluding with O-linked glycosylations (Fig. 2B).

5.1 Epithelial plasticity and N-linked glycosylations in cancer

Lange-Consiglio et al. investigated the glycobiology of the spontaneous transformation of equine amniotic epithelial cells (AECs) into amniotic mesenchymal cells (AMCs), which occurs when the former is grown in culture (Lange-Consiglio, Accogli, Cremonesi, & Desantis, 2014). This transition, referred to as "transdifferentiation" and essentially a form of non-pathologic EMT, was characterized by adoption of an elongated morphology, loss of pan-cytokeratin staining, and gain of vimentin staining. Transdifferentiated cells (termed EMTCs) expressed the mesenchymal stem-cell markers CD29, CD44, CD166, and CD105, as well as MHC-I. Binding of a several lectins was quantified in AECs, amniotic mesenchymal cells (AMCs), and EMTCs to evaluate glycocalyx composition. EMT upregulated expression of highly mannosylated N-linked glycans and O-linked sialoglycans with a terminal NeuNAcα2,3Galβ1,3GalNAc moiety, while downregulating expression of glycans with terminal GlcNAc, fucose, GalNAc, and galactose. Consistent with the non-pathologic nature of transdifferentiation, EMTCs did not lose expression of Siaα2-6Gal/GalNAc as has been observed in metastasis-associated EMT. While these observations collectively support that cells undergoing EMT experience alterations in their glycosylation profile, clear mechanistic explanations for each of these specific glycosylation alterations are not yet available.

Li et al. investigated the role of branched N-glycans in TGF-β1-induced EMT in cancer cells. This group identified the role of Smad signaling in the inhibition of EMT in response to TGF-β1 and in turn the influence of N-GlcNAcylation on this regulatory mechanism (Li et al., 2014). Analysis of patient samples of normal lung epithelium and lung cancers identified lower N-acetylglucosaminyltransferase V (GnT-V) expression in malignancy (except squamous cell carcinoma histology) than in normal tissues, and also that GnT-V expression was positively associated with expression of E-cadherin and negatively associated with N-cadherin and vimentin, suggesting a correlation between GnT-V expression and a stable epithelial phenotype. Analysis of normal lung and lung cancer cell lines identified similar trends. In response to TGF-β1 treatment, known to encourage EMT, A549 cells were found to have lower levels of hybrid and complex N-linked glycans, particularly those with β1,6-GlcNAc branching structures. TGF-β1 treatment significantly reduced GnT-V mRNA and protein expression, while chemical inhibition of β1,6-GlcNAc branched glycan synthesis enhanced the induction of EMT by TGF-β1, as did knockdown of GnT-V by shRNA. Overexpression of GnT-V in A549 cells led to decreased invasiveness and cell migration as well as inhibited EMT in response to TGF-β1, while overexpression of a catalytically inactive GnT-V mutant did not. Further mechanistic investigation found that inhibition of β1,6-GlcNAc branched glycan synthesis or knockdown of GnT-V led to increased Smad2 and Smad3 phosphorylation and nuclear translocation, Smad2/4- and Smad3/4-dependent transcription, and FAK signaling after TGF-β1 treatment. Finally, the overexpression of catalytically active GnT-V led to reciprocal decreases in TGF-β1/Smad signaling. Taken together these results show that TGF-β1 signaling and GnT-V catalysis oppose one another as regulators, positive and negative, respectively, of the EMT in human lung cancer cells.

Khan et al. investigated the role of non-muscle II-A (NMII-A) in a TGF-β1-induced EMT model of NSCLC (Khan et al., 2018). NMII-A is an actin-binding protein that has a central role in cytokinesis, migration and adhesion. In this study, the authors induced EMT in a human NSCLC cell line A549 with TGF-β1, and found decreased expression of NMII-A and E-cadherin as well as increased expression of vimentin at the protein level. They also found increased protein expression of JNK and p-JNK, p-SMAD 2/3, and p-P38; upon genetic knockdown of JNK, SMAD 2/3, and P38, TGF-β1 mediated downregulation of NMII-A was inhibited. The authors hypothesized that TGF-β1 regulates the expression of NMII-A via JNK/P38/PI3K pathway. Genetic knockdown of NMII-A in A549 cells resulted in a mesenchymal cell morphology and decreased proliferation, as well as significantly enhanced migration via wound healing assay and transwell assay. NMII-A knockdown cells had decreased protein expression of epithelial markers E-cadherin and ZO1 and increased expression of mesenchymal markers vimentin, N-cadherin, Slug, Snail and β-catenin. The authors then implanted the NMII-A knockdown cells via tail vein injection into mice, and found that NMII-A knockdown cells disseminated to different organs, including the bones, kidney, liver, ovaries, and lungs, and were found at higher levels in the brain. Overall, the NMII-A knockdown cells had enhanced metastasis to brain and bone compared to control cancer cells. Next, they studied these NMII-A knockdown cells in a mouse lung orthotropic model, and found increased metastasis from the left lung injected with cells to the right lung compared to control. The authors found increased vimentin expression in NMII-A knockdown cells in both lungs, as well as decreased E-cadherin expression in NMII-A knockdown cells in the left lung compared to control. Overall, NMII-A knockdown increased tumor metastasis and modified tumor cell distribution in vivo.

Next, the authors studied the expression of mucin-type core II beta-1,6-N-acetylglucosaminyltransferase (C2GnT-M) in NMII-A knockdown cells. C2GnT-M is located in the Golgi apparatus; when it binds to the

C-terminal region of NMII-A, C2GnT-M is transported to the endoplasmic reticulum, where it is recycled and degraded. In addition, C2GnT-M is responsible for the synthesis of all three branch structures, including core 2, core 4, and I antigen found in the glycans of secreted mucins. These three branch structures are generated by the transfer of GlcNAc from UDP-GlcNAc to core 1, core 3, and I antigen. The authors found increased mRNA and protein levels of C2GnT-M upon NMII-A knockdown. In addition, GnT-V mRNA expression was upregulated and GnT-III was downregulated in two NSCLC cell lines and the NMII-A knockdown cells. When C2GnT-M was additionally knocked down via siRNA in NMII-A knockdown cells, the changes in GnT-V and GnT-III expression were reversed. The authors investigated the role of C2GnT-M in EMT, as NMII-A appeared to regulate the EMT. C2GnT-M knockdown in NMII-A knockdown cells reversed NMII-A knockdown-associated downregulation of E-cadherin and upregulation of vimentin. Knockdown of C2GnT-M alone did not induce a mesenchymal cell morphology; nor did knockdown of NMII-A and C2GnT-M together. In wound healing and transwell assays, knockdown of C2GnT-M in NMII-A knockdown cells deceased cell migration, while inhibition of C2GnT-M alone had no significant effect. C2GnT-M alone did not have an effect on cell motility and morphology. In summary, NMII-A negatively regulates EMT and metastasis via up regulation of C2GnT-M, GnT-V and downregulation of GnT-III.

Mo et al. induced EMT in a human hepatocellular cancer (HCC) or liver cancer cell line (MHCC97-L) with TGF-β1 and found that Smad and Erk1/2 signaling regulated N-acetylglucosaminyltransferase III (GnT-III) expression (Mo et al., 2017). GnT-III, encoded by *MGAT3*, is a glycosyltransferase that catalyzes the synthesis of bisecting GlcNAc structures; GnT-III has been characterized as a suppressor of tumor metastasis. After treating cells with TGF-β1, the authors confirmed the occurrence of EMT in MHCC97-L cells by their observation of spindle-like cellular morphology and decreased expression of E-cadherin and increased expression of genes encoding N-cadherin and α-SMA at the mRNA and protein levels. The authors previously found decreased GnT-III mRNA expression when they induced EMT with hepatocyte growth factor (HGF) in a different human HCC cell line Huh7. In this study, the authors found decreased GnT-III expression at both the mRNA and protein level in their EMT-induced cell line, as well as a reduction in bisecting GlcNAc structure by fluorescence cell PHA-E lectin-immunochemistry. The authors also investigated the Smad-dependent and Erk signaling pathways, as they have been implicated with TGF-β1 signal transduction in EMT. They found slight increased phosphorylation of Smad3 ($<1.5\times$) and Erk1/2 ($>1.5\times$) after TGF-β1 treatment, while the total protein was not significantly changed. Next, they pharmacologically inhibited Smad3 and Erk1/2; SB431542 abolishes Smad3 phosphorylation, and U0126 decreases Erk1/2 phosphorylation. TGF-β1 decreased E-cadherin protein expression; however treatment with either compound restored E-cadherin protein level to untreated level. In addition, treatment with SB431542 or U0126 restored GnT-III mRNA expression as well as bisecting GlcNAc structures. The authors conclude that GnT-III expression is regulated by Smad3 and Erk signaling pathways in this TGF-β1-induced EMT model of HCC.

Cui et al. demonstrated how β1,6-GlcNAc glycans contribute to the role of CD147 function in HCC metastasis (Cui et al., 2018). GnT-V, encoded by *MGAT5* and located in the medial Golgi, catalyzes the addition of β1,6-N-GlcNAc to the alpha-linked mannose of biantennary N-linked oligosaccharides present on newly synthesized glycoproteins. The authors found that GnT-V and EMT were correlated in that GnT-V mediated glycosylation, CD147 protein expression, and CD147-β1,6-branching were elevated during TGF-β1-induced EMT in a normal human hepatic cell line. CD147, also known as basigin, is a tumor-associated transmembrane glycoprotein that carries 1,6-N-acetylglucosamine (1,6-GlcNAc) glycans and in HCC, CD147 is closely associated with EMT, tumorigenesis, and chemoresistance. Next, the authors surveyed the significance of CD147/basigin-β1,6-branched glycans in HCC patients. They analyzed lectin histochemistry of 14 HCC patient samples and observed high levels of CD147/basigin and β1,6-GlcNAc-branched N-glycans. Analysis of 51 patients with HCC revealed that the level of CD147/basigin-β1,6-branched glycans increased with the Barcelona clinic liver cancer (BCLC) stage: patients with stage C disease were markedly positive for CD147/basigin-β1,6-branched glycans, while patients with stage 0-A disease exhibited weak signals. The authors then studied the effect of GnT-V-mediated glycosylation of CD147/basigin on the invasiveness of HCC cells. They inhibited β1,6-branching synthesis with swainsonine in one HCC cell line and found reduced cell invasion in vitro. In addition, they found reduced MMP-1, MMP-2, and MMP-9 mRNA expression when they treated two HCC cell lines with mutant CD147/basigin with defective β1,6-branched N-glycosylation. The authors next assessed the effect of N-glycosylation on the interaction between CD147/basigin and integrin β1. CD147/basigin interacts with integrin β1 to modulate integrin-dependent signaling and focal adhesion kinase (FAK) activation. This interaction activates the Rac/Ras/Raf/ERK and PI3K/Akt pathways and results in enhanced invasive and metastatic potential of HCC cells. Experiments suggested that binding of CD147/basigin to integrin β1 is affected by N-glycosylation. With swainsonine treatment, the ability of CD147/basigin to bind integrin β1 was decreased. Next, the authors showed that pharmacologic inhibition of PI3K in two HCC cell lines resulted in decreased expression of Gnt-V at both the RNA and protein level. Swainsonine treatment impaired PI3K pathway activation and the subsequent activation of p-FAK, p-paxillin, and p-Akt. In summary, the authors found that β1,6-branching promotes cancer cell invasion and CD147-integrin β1 binding in HCC.

In a more descriptive study, Li et al. also induced EMT in a HCC cell line with hepatocyte growth factor (HGF), and analyzed changes in cell surface protein glycosylation with lectin microarray and qRT-PCR (Li et al., 2013). Glycans containing GlcNAc were reduced, and glycans containing β1,6-GlcNAc branching structures were increased. HGF treated cells also had increased invasive potential in transwell migration-invasion assays.

Based on the observation that BGC823 gastric cancer cells had higher levels of GnT-V expression than normal gastric mucosal cells, Huang et al. studied the role of GnT-V in metastatic cellular behavior (Huang et al., 2013). Knockdown of GnT-V with siRNA resulted in decreased N-linked β1,6-branched glycan levels in BGC823 cells. Proliferation rates were decreased with GnT-V knockdown and cells were sensitized to chemical induction of apoptosis. Cells with decreased GnT-V expression showed impaired migration and invasiveness. Both mRNA and protein expression of EGFR, ErbB2, and ErbB4 were decreased by GnT-V knockdown and E-cadherin expression was increased while vimentin and MMP-9 expression were decreased. Overall, knockdown of GnT-V in these cells led to impaired metastatic behavior and a concurrent shift in molecular phenotype toward epithelial and away from mesenchymal characteristics or an MET.

Pinho and colleagues studied the expression of *Mgat3*, the gene encoding GnT-III, and found a correlation between GnT-III and N-linked glycosylation during EP (Pinho et al., 2012). In a normal mouse mammary cell system, induction of EMT with TGF-β1 was associated with decreased mRNA expression of *Mgat3* and significant changes in CpG island methylation patterns within the *Mgat1* gene. Induction of EMT was then found to significantly decrease levels of bisecting GlcNAc structures at the plasma membrane. Bisecting GlcNAc structures were found to co-localize with

E-cadherin in cells with an epithelial phenotype and this co-localization was abrogated by EMT. Immunoprecipitation experiments showed that the epithelial phenotype was associated with E-cadherin modified by bisecting GlcNAc structures, this glycosylation is decreased by EMT, and E-cadherin glycosylation is rescued by MET. Taken together these data suggest that E-cadherin N-glycosylation with bisecting GlcNAc structures is regulated during EP or EMT-MET cycling and associated with GnT-III activity.

A global analysis of N-linked glycan presentation and expression in normal mouse mammary epithelial cells undergoing EMT induced by TGF-β was reported by Tan et al. (2014). Cleavage of N-linked glycans from glycoproteins using PNGase-F was followed by mass spectrometry analysis. As compared to control group cells, those treated with TGF-β displayed a similar number of distinct glycans, but approximately 25% of glycans identified were specific to either control or treated cells, suggesting significant alteration of glycan presentation upon EMT. TGF-β-treated cells expressed more high-mannose N-glycans and fewer complex N-glycans. Treatment with TGF-β resulted in decreased expression of six out of seven N-glycan-related genes, with only *MGAT4B*, the gene encoding GnT-IVb, upregulated. While GnT-IVb catalyzes addition of GlcNAc to complex branching N-glycan structures, the overall decrease in complex N-glycans observed upon TGF-β treatment is thought to result from a significantly greater downregulation of GnT-III as compared to GnT-IV. These findings support a global decrease in N-linked glycans with changes in specific glycan presentation during EMT.

Subsequent work by Tan et al. has implicated loss of GnT-III activity and aberrant glycosylation in promotion of hypoxia-induced EMT in breast cancer cells (Tan, Wang, Li, & Guan, 2018). Exposure of MCF7 and MDA-MB-231 breast cancer cells to hypoxia induced loss of E-cadherin, increased fibronectin, and elevation of glucose transporter-1 (GLUT1), thought to be a marker of hypoxia. Hypoxic cells were more elongated and displayed greater capacity for migration. Mass spectrometry analysis of normoxic and hypoxic cell N-glycans found that most N-glycans specific to hypoxic cells were complex rather than high-mannose. Specifically, bisecting GlcNAc structures decreased with hypoxia, mirroring their earlier findings in TGF-β-induced EMT. Overexpression of GnT-III, responsible for addition of branching GlcNAc moieties to N-linked glycans, inhibited proliferation and migratory capacity in normoxic MCF-7 cells and rescued the hypoxia-inducible E-cadherin downregulation and increased AKT signaling seen in cells with normal GnT-III expression; conversely, knockdown of GnT-III with shRNA enhanced the effects of hypoxia-induced EMT.

Xu and colleagues studied the impact of epidermal growth factor (EGF) on EMT and glycosylation in GE11 epithelial cells (Xu et al., 2017). Treatment with EGF resulted in elongation, decreased intercellular contact, and increased cell migratory ability consistent with EMT. Levels of E-cadherin were decreased and levels of N-cadherin increased, also suggesting EMT induction by EGF. Lectin-binding studied identified significant decreases in bisecting GlcNAc structures and increases in branching GlcNAc structures after EGF treatment. Overexpression of GnT-III in these cells led to resistance to EGF-induced EMT as measured by relative expression of E-cadherin and fibronectin.

A follow-up study by Xu et al. described the glycosylation changes seen in conjunction with changes in cultured cell density (Xu et al., 2017). Plating MCF10A human breast cancer cells at progressively lower densities indicated that sparsely plated cells were more elongated than their more densely plated counterparts. This elongated morphology was accompanied by decreased E-cadherin expression, increased N-cadherin and vimentin, and decreased p120-, α-catenin, and γ-catenin. Taken together the low-density phenotype resembles that seen during EMT. Consistent with this group's reports in an EGF-induced EMT model, low-density cells expressed significantly lower levels of GnT-III and increased GnT-V. Presence of the corresponding glycan structures, bisecting and branched GlcNAc moieties, respectively, was similarly altered. Analysis of β1-integrin N-glycosylation mirrored the overall glycosylation pattern of the cells, with sparsely plated cells displaying higher levels of branched and lower levels of bisecting GlcNAc modifications on β1-integrin.

Hassani et al. found that inhibition of GnT-V, encoded by *MGAT5*, enzymatic activity with the small-molecule inhibitor 3-hydroxy-4,5-bis-benzyloxy-6-benzy-loxymethyl-2-phenyl2-oxo-2l5-[1,2]oxaphosphinane (PST3.1a) restrained glioblastoma multiforme (GBM) growth by affecting microtubule and microfilament integrity of GBM stem cells (Hassani et al., 2017). Overexpression of *MGAT5* is correlated with cell migration, invasion, and EMT; gliomas express highly variable levels of *MGAT5* mRNA, and GnT-V enzymatic activity changes through the course of gliomagenesis. Glycomimetic compounds, phosphinosugars also called phostines, can possibly interfere with glycosylation in cancer cells and the authors selected the lead compound PST3.1a (3-hydroxy-4,5-bis-benzyloxy-6-benzyloxy- methyl-2-phenyl2-oxo-2l5-[1,2]oxaphosphinane) for further anti-tumor studies. PST3.1a inhibited GnT-V enzymatic activity, but displayed no inhibitor effect against GnT-III enzymatic activity. Next, they studied the glycome of GBM cancer stem cells in vitro grown in the presence or absence of PST3.1a, and confirmed that PST3.1a inhibited GnT-V resulting in decreased multi-branched b1,6-GlcNAc mannose intermediate N-glycans. Knocked down *MGAT5* via siRNA was similar to PST3.1a resulting in lower binding of PHA-L lectin. Additional treatment with PST3.1a further reduced residual binding and reduced the ability of GBM cancer stem cells to form large neurospheres (NS) in two GBM cancer stem cell lines in nonadherent/proliferative conditions. The in vivo pharmacologic activity of PST3.1a was then evaluated in orthotropic graft models of GBM using Gli4 and GliT cell lines where they found that PST3.1a treatment significantly reduced the overall Gli4-invaded surface as well as tumor cell densities within invaded areas. Mice treated with PST3.1a survived significantly longer than those treated with vehicle, with a median survival of 108 days versus 79 days for the vehicle cohort. In addition, mice treated with both PST3.1a and temozolomide exhibited a significant survival advantage over those treated with temozolomide alone, with a median survival of 135 days versus 102 days for the vehicle cohort. Finally, using CellMiner and DAVID with the NCI-60 cancer cell line panel, the authors performed transcriptomic analysis and found that mitochondrial gene expression was negatively correlated with PST3.1a cytotoxicity, whereas the expression of genes involved in EMT was positively correlated with the response. In conclusion, PST3.1a inhibits GnT-V, though the anti-tumor effect depends on a cancer cell's EP and energetic metabolic status.

5.2 Epithelial plasticity and O-linked glycosylations in cancer

Ye et al. elucidated a novel mechanism linking mucin-type core 3 O-glycan to the EMT-MET plasticity of colorectal cancer (CRC) cells via a MUC1/p53/miR-200c-dependent signaling cascade (Ye et al., 2017). Mucin-type core 3 O-glycan is one of eight major groups of mucin-type O-glycans; it is synthesized by the enzyme β1,3-N-acetylglucosaminyltransferase-6 (β3gnT6, core 3 synthase), which is primarily expressed in the human stomach, colon and small intestine. Mucin-type core 3 O-glycan had been implicated in cancer metastasis, but the exact mechanism was unclear. The authors first analyzed mRNA expression levels of core 3 synthase in 497 CRC samples in The Cancer Genome Atlas (TCGA). Low expression correlated with lymph node metastasis, distant metastasis, and a poor overall survival for CRC patients. The authors determined that expression levels of core 1 β1,3-galactosyltransferase (C1GalT1); the endoplasmic reticulum chaperone Cosmc involved in C1GalT1 folding; C2GnT1, C2GnT2 and C2GnT3 (three isoforms of core 2 β1,6-N-acetylglucosaminyltransferase (C2GnT)); and core 3 synthase in four human colonic adenocarcinoma cell lines. Low core 3 synthase expression was observed in two of the cell lines (HT-29 and LS174T cells). Core 3 synthase was transfected to re-express core 3 synthase in these cancer cell lines and these β3gnT6/HT-29 and β3gnT6/LS174T recombinant cells exhibited reduced proliferation, migration, and invasion in vitro compared to mock-transfected cells. Next, the authors found that re-expression of core 3 synthase induced a MET as β3gnT6/HT-29 and β3gnT6/LS174T cells were characterized by increased mRNA and protein levels of E-cadherin and decreased mRNA and protein levels of N-cadherin, snail, slug, ZEB1 and ZEB2. Re-expression of core 3 synthase also led to an increase in terminal GlcNAc residues, which were present on mucin-type core 3 O-glycan in β3gnT6 transfected cells; core 3 synthase targeted protein(s) with high-molecular-weight ($>250\,kDa$) and not low-molecular-weight proteins ($<250\,kDa$). In β3gnT6/HT-29 and β3gnT6/LS174T cells core 3 synthase suppressed the translocation of MUC1-C into the nucleus and led to an increase in MUC1-C in the plasma membrane fraction, as revealed by cell fractionation assay. The final mechanism for MET induction by re-expression of core 3 synthase was found to involve a MUC1/p53/miR-200c axis. Inhibition of MUC1 via siRNA, p53 via siRNA, or miR-200c via an inhibitor in β3gnT6/HT-29 and β3gnT6/LS174T cells resulted in increased expression of E-cadherin and decreased expression of N-cadherin, ZEB1 and ZEB2 expression. Using migration and invasion assays, MUC1 knockdown resulted in reduction in migration and invasiveness, while p53 knockdown or miR-200c inhibitor led to an increase in migration and invasiveness. The authors found a positive correlation between core 3 synthase, p53 and miR-200c mRNA levels in 52 CRC samples. Finally, in vivo athymic mice xenograft tumor growth experiments showed the volume and mass of tumors arising from β3gnT6/HT-29 and β3gnT6/LS174T cells were significantly less than those using mock-transfected cells. In summary, the authors concluded that mucin-type core 3 O-glycan is synthesized in the membrane tethered to MUC1 N-terminal domain, which inhibited MUC1-C nucleus translocation, de-repressing p53 gene transcription and finally activating miR-200c expression that lead to a MET.

Based on the totality of these N- and O-linked glycosylation studies only a few consistent associations have been made with regard to the EP programs of EMT and MET. In normal breast epithelial cells and breast cancer GnT-III activity, responsible for addition of branching GlcNAc moieties to N-linked glycans, seems to be negatively associated with an EMT. Apart from this observation, the field is too nascent for any other reliable comments and would generally suggest a very context dependent association between EMT-MET programs and specific glycosylations that calls for further intensive study.

6. Associations between EMT and O-GlcNAcylation

As stated previously, a specific type of O-linked glycosylation, O-GlcNAcylation, is an important posttranslational protein modification regulating diverse cellular processes, such as gene expression, protein stability and cell signaling dynamics. This important biochemical process which is most akin to phosphorylation is controlled by only two highly conserved enzymes: O-linked N-acetylglucosamine transferase (OGT), which transfers N-acetylglucosamine from uridine diphosphate N-acetylglucosamine (UDP-GlcNAc) to protein substrates, and N-acetyl-β-glucosaminidase (OGA, O-GlcNAcase), which removes the O-GlcNAc modification. O-GlcNAcylation regulates practically every cellular process, and plays a major role in the etiology of diseases (Hart, 2014). However, only recently have we begun to appreciate the important roles of O-GlcNAcylation in EMT promoted malignant transformation and tumor progression.

Hints at the connections between O-GlcNAcylation and EMT related phenotypes include a study by Alisson-Silva and colleagues who examined the relationships between the HBP, EMT, and the O-GalNAcylation of human oncofetal fibronectin (onfFN), a mesenchymal extracellular matrix glycoprotein implicated in tumor cell invasiveness and proliferation (Alisson-Silva et al., 2013). They noted that A549 human epithelial lung carcinoma cells exposed to hyperglycemic media displayed increased TGF-β secretion, increased expression of vimentin and N-cadherin, increased motility, and morphological changes consistent with EMT. Hyperglycemia increased expression of UDP-GalNAc:polypeptide N-acetylgalactosaminyltransferase 6 (ppGalNAc-T6), the enzyme responsible for glycosylation of fibronectin (FN) to form onfFN, and this effect was enhanced by but incompletely dependent on TGF-β signaling. Overexpression of GFAT in A549 cells increased expression of the FN IIICS domain which accepts GalNAc from ppGalNAc-T6, as well as total levels of FN, onfFN, N-cadherin, and vimentin. Taken together these data suggest that the routing of glucose through the HBP plays an important role in the initiation of EMT.

This was corroborated by Jang et al. who recently discovered differential O-GlcNAcylation in colorectal carcinoma (CRC) when analyzing primary CRC and nodal metastases (Jang, 2016). Immunohistochemistry was performed on patient tumor samples distinguished as arising from non-neoplastic tissue, centrally within the primary tumor, the invasive front of the primary, or a nodal metastasis. The invasive front was typified by decreased E-cadherin, increased nuclear β-catenin and Snail, and increased O-GlcNAcylation. Central tumor tissue showed intermediate E-cadherin expression, increased O-GlcNAcylation, and levels of nuclear β-catenin and Snail higher than non-neoplastic tissue but lower than the invasive front or nodal metastases. Nodal metastases expressed E-cadherin at levels similar to the tumor center, nuclear β-catenin and Snail similar to the invasive front, and had O-GlcNAcylation levels similar to non-neoplastic tissue. These data suggested a connection between O-GlcNAcylation, invasive behavior, and EMT in human colorectal cancer.

Perhaps these more descriptive data should not have been surprising, as one of the key regulators of EMT–SNAI1 (or Snail or Snail1) was found to be a target of O-GlcNAcylation almost a decade ago. Park and colleagues investigated the stability of the EMT-associated transcription factor Snail1 in the setting of altered O-GlcNAcylation (Park et al., 2010). Serine 112 is a critical regulatory site on Snail1, with phosphorylation at Ser112 resulting in increased degradation. Immunoprecipitation of Snail1 using an O-GlcNAc-specific antibody was followed by mass spectrometry analysis and site-specific mutation experiments allowed localization of Ser112 as a site of O-GlcNAcylation. Chemical inhibition of O-GlcNAcase, resulting in increased O-GlcNAcylation, led to increased Snail1 protein levels at steady state without impacting mRNA expression of Snail1, as well as decreased ubiquitination of Snail1. Overexpression of UDP-GlcNAc–peptide N-acetylglucosaminyltransferase (OGT), the enzyme responsible for O-GlcNAcylation of Snail1, led to decreased Ser112 phosphorylation and increased protein levels of Snail1. Culturing cells in hyperglycemic media increased O-GlcNAcylation and Snail1 protein levels without increasing Snail1 mRNA expression. Hyperglycemia did not, however, increase levels of S112A mutant Snail1 which cannot be O-GlcNAcylated at residue 112. Upregulation of O-GlcNAcylation, both by hyperglycemia and

O-GlcNAcase inhibition, was also found to suppress E-cadherin mRNA transcription, while OGT inhibition led to increased transcription of E-cadherin mRNA. Finally, overexpression of OGT in MCF-7 cells increased cellular migration and invasiveness and was largely rescued with knockdown of Snail1. These findings suggest that O-GlcNAcylation of Snail1, upregulated by increased influx of glucose through the HBP, increases Snail1 stability and induces an EMT.

A more recent study has since confirmed these O-GlcNAcylation results with Snail1. Gao et al. carried out mass spectrometry-based proteomic analysis of proteins that interacted with OGT in HeLa cells and constructed a protein-protein interaction (PPI) network for OGT (Gao et al., 2018). Using co-IP, they confirmed from their mass spectrometry data that OGT physically interacted with the NuRD complex, HDAC1 and HDAC2, which are epigenetic regulators. They hypothesized that one or more components of the NuRD complex may be substrates of O-GlcNAcylation, or that OGT may coordinate with the NuRD complex to participate in histone modification and chromatin remodeling. The authors then focused on elucidating downstream target genes of OGT; they performed RNA-seq on OGT knockdown and control HeLa cells, and identified Snail1 and ING4 as important OGT target genes. Snail1 as mentioned is a canonical EMT-TF, and ING4 is a tumor suppressor that has previously been correlated with high-grade tumors and poor prognosis. The authors also demonstrated that OGT promoted carcinogenesis and metastasis in cervical cancer as genetic knockdown of OGT in two cervical cancer cell lines resulted in delayed cellular migration in an in vitro wound healing assay and a decrease in invasive potential in transwell assay. OGT overexpression in the same cell lines resulted in an increase in invasive potential. Through the Oncomine database, they found that OGT expression was upregulated in cervical cancer compared with adjacent normal tissues. Kaplan–Meier survival analysis of OGT showed that lower expression of OGT was associated with improved survival in cancer patients, when the influence of systemic treatment, endocrine therapy and chemotherapy was excluded. In summary, the authors revealed multiple novel physiological connections between EMT and O-GlcNAcylation using genomic and proteomic analysis.

Harosh-Davidovich et al. have subsequently found that O-GlcNAcylation may enhance cell migration through the regulation of β-catenin and E-cadherin levels in colorectal cancer (CRC) (Harosh-Davidovich & Khalaila, 2018).

The Wnt/β-catenin signaling pathway and cadherin-mediated adhesion are implicated in EMT: β-catenin-mediated transcription upregulates Snail2, which then represses *CDH1* transcription the gene that encodes for E-cadherin. The authors found that with Thiamet-G (TMG), a potent inhibitor of OGA, murine fibroblast cells demonstrated increased O-GlcNAcylation level as well as increased protein levels of β-catenin and E-cadherin. Silencing OGT in the same fibroblast cell line resulted in decreased protein levels of β-catenin. Using WGA affinity purification, they found that β-catenin is O-GlcNAcylated at four sites in fibroblast cells, and TMG treatment resulted in increased β-catenin transcriptional activity. The authors hypothesized that O-GlcNAcylation may prevent the assembly of β-catenin-APC destruction complex, and thus nuclear translocation of β-catenin is enhanced. Fibroblasts treated with TMG exhibited enhanced migration in a wound closure assay. In addition, in an orthotropic murine model of colorectal cancer, shOGT injected mice exhibited lower tumor diameter, lower number of metastasis in the mesentery, and a lower mortality rate. shOGA injected mice displayed an enhanced mortality, while shOGT injected mice had a diminished mortality rate. In summary, increased O-GlcNAcylation enhanced cell migration in vitro, and inhibiting O-GlcNAcylation had anti-tumor effects and reduced metastasis in CRC in vivo.

7. The EMT-HBP-O-GlcNAcylation axis—An important new pathway to promote the neoplastic phenotype

Very recently, a number of studies have shown that EP may not only be impacted by HBP and subsequent O-GlcNAcylation, but that EMT may actually induce HBP-O-GlcNAcylation for malignant transformation and tumor progression constituting an EP-HBP-O-GlcNAcylation axis. Similar to the other glycosylation described above, there appears to be complex regulation of this axis depending on the cellular context in both feed-forward and feed-back modes.

The association of EMT with upregulated HBP genes in cancer has recently been reported in the stromal cells of the lung tumor microenvironment. Zhang et al. performed an integrated analysis of the transcriptome of human primary non-small cell lung cancers along with ^{18}fluoro-2-deoxy-D-glucose PET (FDG-PET) scans and identified *GFPT2* of the HBP pathway to be a critical regulator of metabolic reprogramming in the tumor stroma of lung adenocarcinomas (Zhang et al., 2018). The authors assembled a radiogenomics (RG) study cohort of 130 patients with NSCLC, in which they related clinical glucose uptake via FDG-PET scan to bulk and flow-cell sorted tumor gene expression. The two major histological subtypes, adenocarcinoma and squamous cell carcinoma (SCC), exhibited differences in glucose uptake and gene expression. Genes more highly expressed in adenocarcinoma were enriched for processes related to extracellular matrix (ECM) remodeling and EMT; genes more highly expressed in SCC were enriched for cell growth and proliferation processes. In regard to glucose metabolism, SCC had higher DEGs in glycolysis and pentose phosphate pathway (PPP), supporting the utilization of the Warburg effect, whereas adenocarcinomas showed more DEGs in the HBP, suggesting a role extending beyond the Warburg effect. Using cell-type-specific transcriptomics of the NSCLC tumor microenvironment, the authors found a stronger association of altered glucose uptake in the tumor stroma of adenocarcinoma compared with that of SCC. The authors concluded that metabolic reprogramming as well as the molecular pathways and cell types associated with glucose uptake is histology-specific.

In lung adenocarcinoma, they found that *GFPT2* was the only prognostically significant gene correlated to glucose uptake and that it was predominantly expressed in the tumor stroma. As mentioned above, GFPT2 is a rate-limiting enzyme of the HBP pathway which is also upregulated by the EMT. To validate the role of GFPT2 in EMT, the authors induced EMT in NSCLC adenocarcinoma cells with TGF-β and found increased protein levels of GFPT2 coinciding with decreased E-cadherin and increased vimentin protein levels. Next, the authors induced EMT with TGF-β in normal fibroblasts (NF) derived from normal human lung tissue, and found that these NF transformed into cancer associated fibroblasts (CAFs). In addition, these CAF-like cells expressed HBP genes and EMT genes associated with glucose uptake more highly than NF. At the same time, there was minimal to no change in genes related to glycolysis, PPP, and TCA cycle in these cells. Next, the authors also treated NSCLC adenocarcinoma cell lines (A549, HCC827, and NCI-H358) with TGF-β and found that glucose metabolism genes associated with energy production and cell proliferation (glycolysis, PPP, and the TCA cycle) were mostly unchanged, whereas most EMT genes and several HBP genes correlated directly with increased glucose uptake. In summary, the authors conclude that metabolic reprogramming glucose uptake in tumor stroma correlated strongly with EMT and the HBP.

More direct experimental data linking EMT to the induction of the HBP and increased O-GlcNAcylation come from Lucena et al. who described effects of TGF-β-stimulated EMT in A549 NSCLC cells (Lucena et al., 2016). Treatment of these cells with TGF-β resulted in increased uptake of the fluorescent glucose analogue 2-(N-(7-nitrobenz-2-oxa-1,3-diazol-4-yl)amino)-2-deoxyglucose without changes in lactate, ATP, pyruvate, or glycogen content. Proteomic analysis identified increased protein expression of several enzymes involved in synthesis of the oligosaccharide precursor necessary for protein glycosylation within the ER. Additionally, GFAT1 expression and UDP-GlcNAc production were increased after stimulation of EMT with TGF-β, while glycolytic enzymes downstream of hexokinase as well as pentose phosphate pathway enzymes were stable or downregulated, demonstrating shunting of glucose through the HBP. Characterization of glycoproteins in A549 cells undergoing EMT showed increased glycans with terminal α2-6Neu5Ac groups, poly-N-acetyl-lactosamine, and α-fucose, as well as increased high-mannose N-glycans. Glycans terminating in β-Gal were decreased. Not surprisingly, this glycophenotype differs from that reported in equine EMTCs (Lange-Consiglio et al., 2014), underlining the complexity and specificity of protein glycosylation. Overall, levels of O-linked GlcNAc were increased, as was expression of the enzyme responsible for this modification, OGT. Finally, in the absence of TGF-β stimulation, induction of O-GlcNAcylation was sufficient to induce morphologic, molecular, and behavioral changes consistent with EMT, suggesting that O-GlcNAcylation is an instigating factor in EMT rather than a consequence thereof.

In addition to the links established between EMT-HBP-O-GlcNAcylation in vitro in this one cell line by Lucena et al., Taparra et al. demonstrated that the EMT-HBP axis was capable of accelerating *Kras* mutant lung tumorigenesis, and could be targeted genetically and pharmacologically to produce anti-tumor effects (Taparra et al., 2018). Using genetically engineered mouse models of $Kras^{G12D}$-induced lung tumors, co-overexpression of EMT-TFs *Twist1* or *SNAIL1* was capable of accelerating lung tumorigenesis by inhibition of the tumor suppressor barrier known as oncogene-induced senescence (OIS) (Taparra et al., 2018; Tran et al., 2012). To identify differentially expressed genes (DEGs) in lung tumors derived from these mouse models expressing $Kras^{G12D}$ alone, $Kras^{G12D}/Twist1$, and $Kras^{G12D}/Snail1$, mRNA microarray analysis was performed and revealed genes involved in "metabolic biological processes." Mouse lung tumors overexpressing *Twist1* or *SNAIL1* exhibited increased expression of critical genes in the HBP pathway, specifically *GFPT2* and *UAP1*. Verifying this EMT-HBP axis, overexpression of *SNAIL1* or *Twist1* in human mutant *KRAS*-driven NSCLC cell lines resulted in similar overexpression of HBP genes. Genetic and pharmacologic inhibition of the HBP pathway resulted in cellular senescence in human normal lung cells and NSCLC cells. This premature senescent phenotype was rescued with the bypass GlcN metabolite in normal human lung cells. In addition, in both mouse xenograft and autochthonous lung tumor models, pharmacological inhibition of HBP with DON therapeutically reduced the size of established tumors and inhibited early lung tumorigenesis, respectively, through the induction of senescence and/or cell death. Reversed phase HPLC analysis of human NSCLC cell lines showed that Twist1 overexpression resulted in increased HBP flux via increased levels of UDP-HexNAc (UDP-GlcNAc and downstream metabolite UDP-GalNAc). In addition, OGT protein levels and total O-GlcNAcylation level were increased in both human NSCLC cells and mouse lung tumor tissues that overexpressed EMT-TFs. The HBP pathway was required for $Kras^{G12D}$-induced lung tumorigenesis as OGT deficiency produced by using OGT knockout mice was characterized by a significant delay in lung tumor development. EMT-HBP-O-GlcNAcylation may suppress OIS by stabilizing oncoproteins such as c-Myc which have been shown previously to be O-GlcNAcylated and restrain OIS. Genetic and pharmacological inhibition of O-GlcNAcylation resulted in a reduction of c-Myc protein levels in NSCLC cells in vitro and in vivo. In summary, the EMT-HBP-O-GlcNAcylation axis suppresses cellular senescence and accelerates *Kras* mutant lung tumorigenesis (Fig. 3).

Furthermore, Jiang et al. found that O-GlcNAcylation promotes the metastatic potential of colorectal cancer via a miR-101-O-GlcNAc/EZH2 regulatory feedback circuit (Jiang et al., 2019). TCGA analysis revealed that OGT mRNA level was significantly elevated in all types of CRC, including cecum adenocarcinoma, colon adenocarcinoma, colon mucinous adenocarcinoma, rectal adenocarcinoma, and rectal mucinous adenocarcinoma, compared with that in normal colorectal tissues; there was no significant change in *OGA* mRNA level in CRC. In addition, in 15 cases of CRC patients with lymph node metastases, the expression of O-GlcNAcylation significantly differed between the tumor and lymph node metastases. Next, the authors examined the expression of OGT, OGA, and O-GlcNAcylation in five human CRC cell lines (LoVo, SW620, SW480, HCT-116, and HT-29) and the normal human intestinal epithelial cell line HCoEpiC. They found that O-GlcNAcylation level was significantly increased in all cancer cell lines compared with that in HCoEpiC cells. Overexpression of OGT or treatment with OGA inhibitors PUGNAc

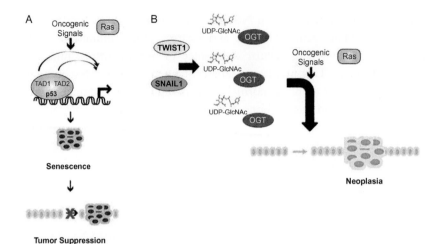

Fig. 3 Epithelial-mesenchymal transition-transcription factors metabolically reprogram normal epithelial cells toward a neoplastic-prone state. (A) Aberrantly activated mutant RAS protein leads to premature cellular senescence of lung epithelial cells as shown by blue cells that is enforced by p53-dependent oncogene-induced senescence. (B) Epithelial-mesenchymal transition-transcription factors (EMT-TFs) TWIST1 and SNAIL1 reprogram epithelial metabolism toward a tumorigenic permissive state by upregulation of the final product of the hexosamine biosynthesis pathway, UDP-N-acetylglucosamine (UDP-GlcNAc), and the O-GlcNAc transferase (OGT)—the enzyme responsible for adding GlcNAc molecules onto proteins. This metabolic shift leads to increased global O-GlcNAcylation that in the background of activated mutant RAS protein results in increased $Kras^{G12D}$-induced lung tumorigenesis.

and TMG increased CRC migration and invasiveness via transwell assays in SW480 cells. In a xenograft murine model using SW480 and SW620 cells, downregulation of O-GlcNAcylation via OGT shRNA in SW620 cells resulted in a significant reduction in the number of metastatic nodules in the liver and lung. When OGT and O-GlcNAcylation were upregulated, the number of metastatic nodules increased.

These authors (Jiang et al., 2019) then implicated the epigenetic factor Enhancer of Zeste Homolog 2 (EZH2) in regulating the EMT in an O-GlcNAcylation–dependent fashion contributing to this increase in metastasis. Downregulation of OGT resulted in decreased expression of mesenchymal markers fibronectin and vimentin, and increased localization of Claudin 7 and E-cadherin on the cell membrane; upregulation of O-GlcNAcylation resulted in the opposite change. Co-IP was performed in SW480 OGT overexpression cells using OGT and O-GlcNAc antibodies, and the co-IP proteins were analyzed by liquid chromatography with tandem mass spectrometry (LC-MS-MS). Among the ~200 proteins observed was histone methyl-transferase EZH2 which is a key epigenetic regulator that catalyzes the tri-methylation of lysine 27 on histone H3 (H3K27me3) and known EMT inducer that participates in the metastasis of various cancers. Treatment with GSK-343, a specific inhibitor of EZH2, inhibited the upregulated invasiveness and reversed the change of EMT markers that was induced by hyper-O-GlcNAcylation. This result was reiterated by genetic knockdown of EZH2 via siRNA.

Next, the authors (Jiang et al., 2019) validated that EZH2 was itself modified by O-GlcNAcylation in SW480 cells. They further tested whether O-GlcNAcylation affected the stability of the EZH2/PRC2 complex. Co-IP experiments showed that the binding capacity of EZH2 and SUZ12, another indispensable component of PRC2, was increased when the O-GlcNAcylation of EZH2 was upregulated by TMG treatment. In addition, O-GlcNAcylation enhanced EZH2 protein stability by prolonging the half-life of degradation and suppressing its ubiquitination. O-GlcNAcylation of EZH2 significantly decreased its ubiquitination level. Next, the authors showed that miR-101 regulates OGT expression in CRC. Using several web-based target prediction algorithms (TargetScanS, miRanda, pictar, and PITA) to identify miRNAs that could potentially target OGT, they identified miR-101 as a potential regulator of OGT expression. This was confirmed by a luciferase reporter assay, in which the luciferase activity of the reporter constructs containing OGT 3′-untranslated region (UTR) was most significantly reduced by a miR-101 mimic construct, as well as a "rescue" experiment using miR-101 to target a mutated seed sequence of miR-101 at the OGT 3′-UTR. Examining the miR-101 expression levels in human CRC cell lines, the authors found that miR-101 was significantly decreased in all cancer cell lines compared with that in the HCoEpiC cell line. When SW620 cells were transfected with the miR-101 mimic, OGT and EZH2 protein levels decreased; SW480 cells similarly transfected had increased OGT and EZH2 levels. In addition, SW620 cells transfected with miR-101 mimics had a significant decrease in migratory and invasive abilities; silencing miR-101 in the SW480 cells with an antisense oligonucleotide inhibitor significantly increased cell migration and invasion. Next, the authors showed that miR-101 regulates EMT by targeting OGT and EZH2 such that transfection with miR-101 resulted in decreased protein level of fibronectin and O-GlcNAcylation levels, as well as increased level of E-cadherin in SW620 cell. The opposite effects were seen in SW480 cells with a miR-101 inhibitor. In addition, overexpression of OGT or EZH2 in the SW480 cells reversed miR-101-mediated decrease in fibronectin protein expression and increase of E-cadherin protein expression.

Finally, Jiang et al. (2019) hypothesized that OGT feedback may transcriptionally silence miR-101 in an EZH2-dependent manner. They found that when OGT or EZH2 were downregulated, the levels of mature miR-101, precursor miR-101-1, and precursor miR-101-2 were increased; however if EZH2 was silenced in advance, there was no significant change when OGT was downregulated. Results from ChIP-qPCR analysis of the transcriptional start regions (TSSs) of miR-101-1 and miR-101-2 precursors verified that the miR-101 promoter regions were highly enriched in EZH2, H3K27me3, and O-GlcNAcylation. In addition, EZH2 knockdown in SW480 cells almost completely eliminated the enrichment of O-GlcNAcylation and H3K27me3 in the mIR-101 TSS region. When O-GlcNAcylation was upregulated, the H3K27me3 enrichment in these regions was significantly increased. Meanwhile, the knockdown of EZH2 in SW620 eliminated the enrichment of O-GlcNAcylation in the miR-101 TSS regions, while the knockdown of OGT in the same cell line eliminated the enrichment of H3K27me3 in this region. These results support the hypothesis that O-GlcNAcylation feedback negatively regulates miR-101 and that EZH2 is a necessary component of this loop. In summary, in CRC cells, miR-101/O-GlcNAcylation/EZH2 signaling forms a feedback loop that promotes EMT-mediated metastasis. The downregulation of miR-101 in CRC promotes the elevation of O-GlcNAcylation and thus enhances EZH2 protein stability and function, which, in turn, further reduces the expression of miR-101.

8. Conclusion and future directions

In summary, EP drives metabolic reprograming and facilitates a number of context-specific, complex glycosylation events. Depending on the context EP programs are associated with intricate changes, both positive and negative in nature, to extracellular and intracellular glycoconjugates that are associated with the neoplastic phenotype. In this chapter, we also highlight the emerging knowledge of an EP-mediated glycosylation event, the redirection of increased glucose flux toward the HBP resulting in increased global cellular O-GlcNAcylation, which recent data suggest can contribute to tumorigenesis and cancer progression. Although the HBP-O-GlcNAcylation axis is emerging as an important pathway in the cancer cell during EMT, the precise molecular mechanisms of O-GlcNAcylation modification during EMT promoted tumorigenesis and metastasis remain to be elucidated. Further investigations are needed to identify novel O-GlcNAcylated proteins involved in signaling pathways leading to downstream execution of EP programs. It is also likely that the role of EMT-HBP-O-GlcNAcylation discussed here may extend beyond cancer development and metastasis to include other known EMT-related phenotypes such as cancer treatment resistance to radiation, cytotoxic systemic agents, targeted therapies and immunotherapies (Arumugam et al., 2009; Efimova et al., 2016; Shintani et al., 2011; Yochum et al., 2019; Zheng et al., 2015). In addition, although EP programs are highly conserved developmental programs that have been studied for decades, the requirement of this EMT-HBP-O-GlcNAcylation axis in normal development is unknown and deserves analysis. Finally, the HBP and O-GlcNAcylation present themselves as important potential novel targetable pathways to target EP in cancers.

Acknowledgments

R.M.P. was supported by the RSNA. H.W. was supported by the Johns Hopkins-Allegheny Health Network Research Fund. P.T.T. was supported by the Nesbitt-McMaster Foundation, Ronald Rose & Joan Lazar; Movember Foundation, Prostate Cancer Foundation; NIH/NCI (R01CA166348, U01CA212007, U01CA231776 and R21CA223403).

Conflict of interest

The authors confirm no conflict of interest.

References

Aebi, M. (2013). N-linked protein glycosylation in the ER. *Biochimica et Biophysica Acta*, *1833*(11), 2430–2437. https://doi.org/10.1016/j.bbamcr.2013.04.001.

Alisson-Silva, F., Freire-de-Lima, L., Donadio, J. L., Lucena, M. C., Penha, L., Sa-Diniz, J. N., et al. (2013). Increase of O-glycosylated oncofetal fibronectin in high glucose-induced epithelial-mesenchymal transition of cultured human epithelial cells. *PLoS One*, *8*(4). e60471. https://doi.org/10.1371/journal.pone.0060471.

Ansieau, S., Bastid, J., Doreau, A., Morel, A. P., Bouchet, B. P., Thomas, C., et al. (2008). Induction of EMT by twist proteins as a collateral effect of tumor-promoting inactivation of premature senescence. *Cancer Cell*, *14*(1), 79–89. https://doi.org/10.1016/j.ccr.2008.06.005.

Arumugam, T., Ramachandran, V., Fournier, K. F., Wang, H., Marquis, L., Abbruzzese, J. L., et al. (2009). Epithelial to mesenchymal transition contributes to drug resistance in pancreatic cancer. *Cancer Research*, *69*(14), 5820–5828. https://doi.org/10.1158/0008-5472.CAN-08-2819.

Boehmelt, G., Fialka, I., Brothers, G., McGinley, M. D., Patterson, S. D., Mo, R., et al. (2000). Cloning and characterization of the murine glucosamine-6-phosphate acetyltransferase EMeg32. Differential expression and intracellular membrane association. *The Journal of Biological Chemistry, 275*(17), 12821–12832.

Boehmelt, G., Wakeham, A., Elia, A., Sasaki, T., Plyte, S., Potter, J., et al. (2000). Decreased UDP-GlcNAc levels abrogate proliferation control in EMeg32-deficient cells. *The EMBO Journal, 19*(19), 5092–5104. https://doi.org/10.1093/emboj/19.19.5092.

Brabletz, S., Brabletz, T., & Stemmler, M. P. (2017). Road to perdition: Zeb1-dependent and -independent ways to metastasis. *Cell Cycle, 16*(19), 1729–1730. https://doi.org/10.1080/15384101.2017.1360648.

Brockhausen, I., Schachter, H., & Stanley, P. (2009). O-GalNAc glycans. In A. Varki, R. D. Cummings, J. D. Esko, H. H. Freeze, P. Stanley, C. R. Bertozzi, G. W. Hart, & M. E. Etzler (Eds.), *Essentials of glycobiology*. Cold Spring Harbor, NY: Cold Spring Harbor Laboratory Press.

Burns, T. F., Dobromilskaya, I., Murphy, S. C., Gajula, R. P., Thiyagarajan, S., Chatley, S. N., et al. (2013). Inhibition of TWIST1 leads to activation of oncogene-induced senescence in oncogene-driven non-small cell lung cancer. *Molecular Cancer Research, 11*(4), 329–338. https://doi.org/10.1158/1541-7786.

Carvalho-Cruz, P., Alisson-Silva, F., Todeschini, A. R., & Dias, W. B. (2018). Cellular glycosylation senses metabolic changes and modulates cell plasticity during epithelial to mesenchymal transition. *Developmental Dynamics, 247*(3), 481–491. https://doi.org/10.1002/dvdy.24553.

Chaffer, C. L., & Weinberg, R. A. (2011). A perspective on cancer cell metastasis. *Science, 331*(6024), 1559–1564. https://doi.org/10.1126/science.1203543.

Chen, Y., LeBleu, V. S., Carstens, J. L., Sugimoto, H., Zheng, X., Malasi, S., et al. (2018). Dual reporter genetic mouse models of pancreatic cancer identify an epithelial-to-mesenchymal transition-independent metastasis program. *EMBO Molecular Medicine, 10*(10). https://doi.org/10.15252/emmm.201809085. pii: e9085.

Costa, A. R., Rodrigues, M. E., Henriques, M., Oliveira, R., & Azeredo, J. (2014). Glycosylation: Impact, control and improvement during therapeutic protein production. *Critical Reviews in Biotechnology, 34*(4), 281–299. https://doi.org/10.3109/07388551.2013.793649.

Cui, J., Huang, W., Wu, B., Jin, J., Jing, L., Shi, W. P., et al. (2018). N-glycosylation by N-acetylglucosaminyltransferase V enhances the interaction of CD147/basigin with integrin beta1 and promotes HCC metastasis. *The Journal of Pathology, 245*(1), 41–52. https://doi.org/10.1002/path.5054.

Dennis, J. W., Lau, K. S., Demetriou, M., & Nabi, I. R. (2009). Adaptive regulation at the cell surface by N-glycosylation. *Traffic, 10*(11), 1569–1578. https://doi.org/10.1111/j.1600-0854.2009.00981.x.

Dong, C., Yuan, T., Wu, Y., Wang, Y., Fan, T. W., Miriyala, S., et al. (2013). Loss of FBP1 by Snail-mediated repression provides metabolic advantages in basal-like breast cancer. *Cancer Cell, 23*(3), 316–331. https://doi.org/10.1016/j.ccr.2013.01.022.

Dongre, A., Rashidian, M., Reinhardt, F., Bagnato, A., Keckesova, Z., Ploegh, H. L., et al. (2017). Epithelial-to-mesenchymal transition contributes to immunosuppression in breast carcinomas. *Cancer Research, 77*(15), 3982–3989. https://doi.org/10.1158/0008-5472.CAN-16-3292.

Duan, F., Jia, D., Zhao, J., Wu, W., Min, L., Song, S., et al. (2016). Loss of GFAT1 promotes epithelial-to-mesenchymal transition and predicts unfavorable prognosis in gastric cancer. *Oncotarget, 7*(25), 38427–38439. https://doi.org/10.18632/oncotarget.9538.

Durand, P., Golinelli-Pimpaneau, B., Mouilleron, S., Badet, B., & Badet-Denisot, M. A. (2008). Highlights of glucosamine-6P synthase catalysis. *Archives of Biochemistry and Biophysics, 474*(2), 302–317. https://doi.org/10.1016/j.abb.2008.01.026.

Efimova, E. V., Takahashi, S., Shamsi, N. A., Wu, D., Labay, E., Ulanovskaya, O. A., et al. (2016). Linking cancer metabolism to DNA repair and accelerated senescence. *Molecular*

Cancer Research, 14(2), 173–184. https://doi.org/10.1158/1541-7786.MCR-15-02631541-7786.MCR-15-0263 [pii].

Fischer, K. R., Durrans, A., Lee, S., Sheng, J., Li, F., Wong, S. T., et al. (2015). Epithelial-to-mesenchymal transition is not required for lung metastasis but contributes to chemoresistance. Nature, 527(7579), 472–476. https://doi.org/10.1038/nature15748.

Gao, J., Yang, Y., Qiu, R., Zhang, K., Teng, X., Liu, R., et al. (2018). Proteomic analysis of the OGT interactome: Novel links to epithelial-mesenchymal transition and metastasis of cervical cancer. Carcinogenesis, 39(10), 1222–1234. https://doi.org/10.1093/carcin/bgy097.

Gordan, J. D., Thompson, C. B., & Simon, M. C. (2007). HIF and c-Myc: Sibling rivals for control of cancer cell metabolism and proliferation. Cancer Cell, 12(2), 108–113.

Guo, H. B., Lee, I., Kamar, M., & Pierce, M. (2003). N-acetylglucosaminyltransferase V expression levels regulate cadherin-associated homotypic cell-cell adhesion and intracellular signaling pathways. The Journal of Biological Chemistry, 278(52), 52412–52424. https://doi.org/10.1074/jbc.M308837200M308837200 [pii].

Guppy, M., Greiner, E., & Brand, K. (1993). The role of the Crabtree effect and an endogenous fuel in the energy metabolism of resting and proliferating thymocytes. European Journal of Biochemistry, 212(1), 95–99.

Harosh-Davidovich, S. B., & Khalaila, I. (2018). O-GlcNAcylation affects beta-catenin and E-cadherin expression, cell motility and tumorigenicity of colorectal cancer. Experimental Cell Research, 364(1), 42–49. https://doi.org/10.1016/j.yexcr.2018.01.024.

Hart, G. W. (2014). Minireview series on the thirtieth anniversary of research on O-GlcNAcylation of nuclear and cytoplasmic proteins: Nutrient regulation of cellular metabolism and physiology by O-GlcNAcylation. The Journal of Biological Chemistry, 289(50), 34422–34423. https://doi.org/10.1074/jbc.R114.609776.

Hart, G. W., & Akimoto, Y. (2009). The O-GlcNAc modification. In A. Varki, R. D. Cummings, J. D. Esko, H. H. Freeze, P. Stanley, C. R. Bertozzi, G. W. Hart, & M. E. Etzler (Eds.), Essentials of glycobiology: Cold Spring Harbor Laboratory Press, Cold Spring Harbor, NY.

Hassani, Z., Saleh, A., Turpault, S., Khiati, S., Morelle, W., Vignon, J., et al. (2017). Phostine PST3.1a targets MGAT5 and inhibits glioblastoma-initiating cell invasiveness and proliferation. Molecular Cancer Research, 15(10), 1376–1387. https://doi.org/10.1158/1541-7786.MCR-17-0120.

Huang, B., Sun, L., Cao, J., Zhang, Y., Wu, Q., Zhang, J., et al. (2013). Downregulation of the GnT-V gene inhibits metastasis and invasion of BGC823 gastric cancer cells. Oncology Reports, 29(6), 2392–2400. https://doi.org/10.3892/or.2013.2373.

Itkonen, H. M., Engedal, N., Babaie, E., Luhr, M., Guldvik, I. J., Minner, S., et al. (2015). UAP1 is overexpressed in prostate cancer and is protective against inhibitors of N-linked glycosylation. Oncogene, 34(28), 3744–3750. https://doi.org/10.1038/onc.2014.307.

Iyer, S. P., & Hart, G. W. (2003). Dynamic nuclear and cytoplasmic glycosylation: Enzymes of O-GlcNAc cycling. Biochemistry, 42(9), 2493–2499. https://doi.org/10.1021/bi020685a.

Jang, T. J. (2016). Differential membranous E-cadherin expression, cell proliferation and O-GlcNAcylation between primary and metastatic nodal lesion in colorectal cancer. Pathology, Research and Practice, 212(2), 113–119. https://doi.org/10.1016/j.prp.2015.12.003.

Jia, D., Jolly, M. K., Boareto, M., Parsana, P., Mooney, S. M., Pienta, K. J., et al. (2015). OVOL guides the epithelial-hybrid-mesenchymal transition. Oncotarget, 6(17), 15436–15448. https://doi.org/10.18632/oncotarget.3623.

Jiang, M., Xu, B., Li, X., Shang, Y., Chu, Y., Wang, W., et al. (2019). O-GlcNAcylation promotes colorectal cancer metastasis via the miR-101-O-GlcNAc/EZH2 regulatory feedback circuit. Oncogene, 38(3), 301–316. https://doi.org/10.1038/s41388-018-0435-5.

Jones, R. G., & Thompson, C. B. (2009). Tumor suppressors and cell metabolism: A recipe for cancer growth. *Genes & Development, 23*(5), 537–548. https://doi.org/10.1101/gad.1756509.

Khan, G. J., Gao, Y., Gu, M., Wang, L., Khan, S., Naeem, F., et al. (2018). TGF-beta1 causes EMT by regulating N-acetyl glucosaminyl transferases via downregulation of non muscle myosin II-A through JNK/P38/PI3K pathway in lung cancer. *Current Cancer Drug Targets, 18*(2), 209–219. https://doi.org/10.2174/1568009617666170807120304.

Kondaveeti, Y., Guttilla Reed, I. K., & White, B. A. (2015). Epithelial-mesenchymal transition induces similar metabolic alterations in two independent breast cancer cell lines. *Cancer Letters, 364*(1), 44–58. https://doi.org/10.1016/j.canlet.2015.04.025.

Kowalik, M. A., Columbano, A., & Perra, A. (2017). Emerging role of the pentose phosphate pathway in hepatocellular carcinoma. *Frontiers in Oncology, 7,* 87. https://doi.org/10.3389/fonc.2017.00087.

Krebs, A. M., Mitschke, J., Lasierra Losada, M., Schmalhofer, O., Boerries, M., Busch, H., et al. (2017). The EMT-activator Zeb1 is a key factor for cell plasticity and promotes metastasis in pancreatic cancer. *Nature Cell Biology, 19*(5), 518–529. https://doi.org/10.1038/ncb3513ncb3513.

Labak, C. M., Wang, P. Y., Arora, R., Guda, M. R., Asuthkar, S., Tsung, A. J., et al. (2016). Glucose transport: Meeting the metabolic demands of cancer, and applications in glioblastoma treatment. *American Journal of Cancer Research, 6*(8), 1599–1608.

Lange-Consiglio, A., Accogli, G., Cremonesi, F., & Desantis, S. (2014). Cell surface glycan changes in the spontaneous epithelial-mesenchymal transition of equine amniotic multipotent progenitor cells. *Cells, Tissues, Organs, 200*(3–4), 212–226. https://doi.org/10.1159/000433420.

Lau, K. S., Partridge, E. A., Grigorian, A., Silvescu, C. I., Reinhold, V. N., Demetriou, M., et al. (2007). Complex N-glycan number and degree of branching cooperate to regulate cell proliferation and differentiation. *Cell, 129*(1), 123–134. https://doi.org/10.1016/j.cell.2007.01.049.

Lee, K. E., & Bar-Sagi, D. (2010). Oncogenic KRas suppresses inflammation-associated senescence of pancreatic ductal cells. *Cancer Cell, 18*(5), 448–458. https://doi.org/10.1016/j.ccr.2010.10.020S1535-6108(10)00422-8 [pii].

Leithner, K., Hrzenjak, A., Trotzmuller, M., Moustafa, T., Kofeler, H. C., Wohlkoenig, C., et al. (2015). PCK2 activation mediates an adaptive response to glucose depletion in lung cancer. *Oncogene, 34*(8), 1044–1050. https://doi.org/10.1038/onc.2014.47.

Li, S., Mo, C., Peng, Q., Kang, X., Sun, C., Jiang, K., et al. (2013). Cell surface glycan alterations in epithelial mesenchymal transition process of Huh7 hepatocellular carcinoma cell. *PLoS One, 8*(8). e71273https://doi.org/10.1371/journal.pone.0071273.

Li, C., Rodriguez, M., & Banerjee, D. (2000). Cloning and characterization of complementary DNA encoding human N-acetylglucosamine-phosphate mutase protein. *Gene, 242*(1–2), 97–103.

Li, N., Xu, H., Fan, K., Liu, X., Qi, J., Zhao, C., et al. (2014). Altered beta1,6-GlcNAc branched N-glycans impair TGF-beta-mediated epithelial-to-mesenchymal transition through Smad signalling pathway in human lung cancer. *Journal of Cellular and Molecular Medicine, 18*(10), 1975–1991. https://doi.org/10.1111/jcmm.12331.

Li, S., & Yang, J. (2014). Ovol proteins: Guardians against EMT during epithelial differentiation. *Developmental Cell, 29*(1), 1–2. https://doi.org/10.1016/j.devcel.2014.04.002.

Liu, Q., Luo, Q., Halim, A., & Song, G. (2017). Targeting lipid metabolism of cancer cells: A promising therapeutic strategy for cancer. *Cancer Letters, 401,* 39–45. https://doi.org/10.1016/j.canlet.2017.05.002.

Liu, M., Quek, L. E., Sultani, G., & Turner, N. (2016). Epithelial-mesenchymal transition induction is associated with augmented glucose uptake and lactate production in pancreatic ductal adenocarcinoma. *Cancer & Metabolism, 4,* 19. https://doi.org/10.1186/s40170-016-0160-x.

Loibl, M., & Strahl, S. (2013). Protein O-mannosylation: What we have learned from baker's yeast. *Biochimica et Biophysica Acta, 1833*(11), 2438–2446. https://doi.org/10.1016/j.bbamcr.2013.02.008.

Lucena, M. C., Carvalho-Cruz, P., Donadio, J. L., Oliveira, I. A., de Queiroz, R. M., Marinho-Carvalho, M. M., et al. (2016). Epithelial mesenchymal transition induces aberrant glycosylation through hexosamine biosynthetic pathway activation. *The Journal of Biological Chemistry, 291*(25), 12917–12929. https://doi.org/10.1074/jbc.M116.729236.

Malek, R., Wang, H., Taparra, K., & Tran, P. T. (2017). Therapeutic targeting of epithelial plasticity programs: Focus on the epithelial-mesenchymal transition. *Cells, Tissues, Organs, 203*(2), 114–127. https://doi.org/10.1159/000447238000447238.

Mani, S. A., Guo, W., Liao, M. J., Eaton, E. N., Ayyanan, A., Zhou, A. Y., et al. (2008). The epithelial-mesenchymal transition generates cells with properties of stem cells. *Cell, 133*(4), 704–715. https://doi.org/10.1016/j.cell.2008.03.027.

Marshall, S., Bacote, V., & Traxinger, R. R. (1991). Complete inhibition of glucose-induced desensitization of the glucose transport system by inhibitors of mRNA synthesis. Evidence for rapid turnover of glutamine: Fructose-6-phosphate amidotransferase. *The Journal of Biological Chemistry, 266*(16), 10155–10161.

Mo, C., Liu, T., Zhang, S., Guo, K., Li, M., Qin, X., et al. (2017). Reduced N-acetylglucosaminyltransferase III expression via Smad3 and Erk signaling in TGF-beta1-induced HCC EMT model. *Discovery Medicine, 23*(124), 7–17.

Morandi, A., Taddei, M. L., Chiarugi, P., & Giannoni, E. (2017). Targeting the metabolic reprogramming that controls epithelial-to-mesenchymal transition in aggressive tumors. *Frontiers in Oncology, 7*, 40. https://doi.org/10.3389/fonc.2017.00040.

Morel, A. P., Ginestier, C., Pommier, R. M., Cabaud, O., Ruiz, E., Wicinski, J., et al. (2017). A stemness-related ZEB1-MSRB3 axis governs cellular pliancy and breast cancer genome stability. *Nature Medicine, 23*(5), 568–578. https://doi.org/10.1038/nm.4323nm.4323.

Ni, T., Li, X. Y., Lu, N., An, T., Liu, Z. P., Fu, R., et al. (2016). Snail1-dependent p53 repression regulates expansion and activity of tumour-initiating cells in breast cancer. *Nature Cell Biology, 18*(11), 1221–1232. https://doi.org/10.1038/ncb3425ncb3425.

Nieto, M. A., Huang, R. Y., Jackson, R. A., & Thiery, J. P. (2016). Emt: 2016. *Cell, 166*(1), 21–45. https://doi.org/10.1016/j.cell.2016.06.028.

Osthus, R. C., Shim, H., Kim, S., Li, Q., Reddy, R., Mukherjee, M., et al. (2000). Deregulation of glucose transporter 1 and glycolytic gene expression by c-Myc. *The Journal of biological chemistry, 275*(29), 21797–21800.

Park, S. Y., Kim, H. S., Kim, N. H., Ji, S., Cha, S. Y., Kang, J. G., et al. (2010). Snail1 is stabilized by O-GlcNAc modification in hyperglycaemic condition. *The EMBO Journal, 29*(22), 3787–3796. https://doi.org/10.1038/emboj.2010.254.

Peneff, C., Ferrari, P., Charrier, V., Taburet, Y., Monnier, C., Zamboni, V., et al. (2001). Crystal structures of two human pyrophosphorylase isoforms in complexes with UDPGlc(Gal)NAc: Role of the alternatively spliced insert in the enzyme oligomeric assembly and active site architecture. *The EMBO Journal, 20*(22), 6191–6202. https://doi.org/10.1093/emboj/20.22.6191.

Pinho, S. S., Oliveira, P., Cabral, J., Carvalho, S., Huntsman, D., Gartner, F., et al. (2012). Loss and recovery of Mgat3 and GnT-III mediated E-cadherin N-glycosylation is a mechanism involved in epithelial-mesenchymal-epithelial transitions. *PLoS One, 7*(3). e33191https://doi.org/10.1371/journal.pone.0033191.

Pinho, S. S., Osorio, H., Nita-Lazar, M., Gomes, J., Lopes, C., Gartner, F., et al. (2009). Role of E-cadherin N-glycosylation profile in a mammary tumor model. *Biochemical and Biophysical Research Communications, 379*(4), 1091–1096. https://doi.org/10.1016/j.bbrc.2009.01.024.

Puisieux, A., Brabletz, T., & Caramel, J. (2014). Oncogenic roles of EMT-inducing transcription factors. *Nature Cell Biology*, *16*(6), 488–494. https://doi.org/10.1038/ncb2976ncb2976.

Puisieux, A., Pommier, R. M., Morel, A. P., & Lavial, F. (2018). Cellular pliancy and the multistep process of tumorigenesis. *Cancer Cell*, *33*(2), 164–172. https://doi.org/10.1016/j.ccell.2018.01.007.

Rao, X., Duan, X., Mao, W., Li, X., Li, Z., Li, Q., et al. (2015). O-GlcNAcylation of G6PD promotes the pentose phosphate pathway and tumor growth. *Nature Communications*, *6*, 8468. https://doi.org/10.1038/ncomms9468.

Ricciardiello, F., Votta, G., Palorini, R., Raccagni, I., Brunelli, L., Paiotta, A., et al. (2018). Inhibition of the hexosamine biosynthetic pathway by targeting PGM3 causes breast cancer growth arrest and apoptosis. *Cell Death & Disease*, *9*(3), 377. https://doi.org/10.1038/s41419-018-0405-4.

Roca, H., Hernandez, J., Weidner, S., McEachin, R. C., Fuller, D., Sud, S., et al. (2013). Transcription factors OVOL1 and OVOL2 induce the mesenchymal to epithelial transition in human cancer. *PLoS One*, *8*(10). e76773https://doi.org/10.1371/journal.pone.0076773.

Sasai, K., Ikeda, Y., Fujii, T., Tsuda, T., & Taniguchi, N. (2002). UDP-GlcNAc concentration is an important factor in the biosynthesis of beta1,6-branched oligosaccharides: Regulation based on the kinetic properties of N-acetylglucosaminyltransferase V. *Glycobiology*, *12*(2), 119–127.

Shaul, Y. D., Freinkman, E., Comb, W. C., Cantor, J. R., Tam, W. L., Thiru, P., et al. (2014). Dihydropyrimidine accumulation is required for the epithelial-mesenchymal transition. *Cell*, *158*(5), 1094–1109. https://doi.org/10.1016/j.cell.2014.07.032.

Shintani, Y., Okimura, A., Sato, K., Nakagiri, T., Kadota, Y., Inoue, M., et al. (2011). Epithelial to mesenchymal transition is a determinant of sensitivity to chemoradiotherapy in non-small cell lung cancer. *The Annals of Thoracic Surgery*, *92*(5), 1794–1804. discussion 1804. https://doi.org/10.1016/j.athoracsur.2011.07.032.

Shiota, M., Izumi, H., Onitsuka, T., Miyamoto, N., Kashiwagi, E., Kidani, A., et al. (2008). Twist and p53 reciprocally regulate target genes via direct interaction. *Oncogene*, *27*(42), 5543–5553. https://doi.org/10.1038/onc.2008.176onc2008176.

Tan, Z., Lu, W., Li, X., Yang, G., Guo, J., Yu, H., et al. (2014). Altered N-Glycan expression profile in epithelial-to-mesenchymal transition of NMuMG cells revealed by an integrated strategy using mass spectrometry and glycogene and lectin microarray analysis. *Journal of Proteome Research*, *13*(6), 2783–2795. https://doi.org/10.1021/pr401185z.

Tan, Z., Wang, C., Li, X., & Guan, F. (2018). Bisecting N-acetylglucosamine structures inhibit hypoxia-induced epithelial-mesenchymal transition in breast cancer cells. *Frontiers in Physiology*, *9*, 210. https://doi.org/10.3389/fphys.2018.00210.

Taparra, K., Tran, P. T., & Zachara, N. E. (2016). Hijacking the hexosamine biosynthetic pathway to promote EMT-mediated neoplastic phenotypes. *Frontiers in Oncology*, *6*, 85. https://doi.org/10.3389/fonc.2016.00085.

Taparra, K., Wang, H., Malek, R., Lafargue, A., Barbhuiya, M. A., Wang, X., et al. (2018). O-GlcNAcylation is required for mutant KRAS-induced lung tumorigenesis. *The Journal of Clinical Investigation*, *128*(11), 4924–4937. https://doi.org/10.1172/JCI94844.

Thomson, S., Petti, F., Sujka-Kwok, I., Epstein, D., & Haley, J. D. (2008). Kinase switching in mesenchymal-like non-small cell lung cancer lines contributes to EGFR inhibitor resistance through pathway redundancy. *Clinical & Experimental Metastasis*, *25*(8), 843–854. https://doi.org/10.1007/s10585-008-9200-4.

Tran, P. T., Shroff, E. H., Burns, T. F., Thiyagarajan, S., Das, S. T., Zabuawala, T., et al. (2012). Twist1 suppresses senescence programs and thereby accelerates and maintains mutant Kras-induced lung tumorigenesis. *PLoS Genetics*, *8*(5). e1002650https://doi.org/10.1371/journal.pgen.1002650.

Tsai, J. H., Donaher, J. L., Murphy, D. A., Chau, S., & Yang, J. (2012). Spatiotemporal regulation of epithelial-mesenchymal transition is essential for squamous cell carcinoma metastasis. *Cancer Cell, 22*(6), 725–736. https://doi.org/10.1016/j.ccr.2012.09.022.

Vander Heiden, M. G., Cantley, L. C., & Thompson, C. B. (2009). Understanding the Warburg effect: The metabolic requirements of cell proliferation. *Science, 324*(5930), 1029–1033. https://doi.org/10.1126/science.1160809.

Varki, A., Kannagi, R., Toole, B., & Stanley, P. (2015). Glycosylation changes in cancer. In A. Varki, R. D. Cummings, J. D. Esko, P. Stanley, G. W. Hart, M. Aebi, A. G. Darvill, T. Kinoshita, N. H. Packer, J. H. Prestegard, R. L. Schnaar, & P. H. Seeberger (Eds.), *Essentials of glycobiology* (pp. 597–609): Cold Spring Harbor Laboratory Press. Cold Spring Harbor, NY.

Wang, P., Wang, H., Gai, J., Tian, X., Zhang, X., Lv, Y., et al. (2017). Evolution of protein N-glycosylation process in Golgi apparatus which shapes diversity of protein N-glycan structures in plants, animals and fungi. *Scientific Reports, 7*, 40301. https://doi.org/10.1038/srep40301.

Wang-Gillam, A., Pastuszak, I., & Elbein, A. D. (1998). A 17-amino acid insert changes UDP-N-acetylhexosamine pyrophosphorylase specificity from UDP-GalNAc to UDP-GlcNAc. *The Journal of Biological Chemistry, 273*(42), 27055–27057.

Warburg, O. (1925). The metabolism of carcinoma cells. *The Journal of Cancer Research, 9*(1), 148–163. https://doi.org/10.1158/jcr.1925.148.

Warburg, O., Wind, F., & Negelein, E. (1927). The metabolism of tumors in the body. *The Journal of General Physiology, 8*(6), 519–530.

Weigert, C., Friess, U., Brodbeck, K., Haring, H. U., & Schleicher, E. D. (2003). Glutamine: Fructose-6-phosphate aminotransferase enzyme activity is necessary for the induction of TGF-beta1 and fibronectin expression in mesangial cells. *Diabetologia, 46*(6), 852–855. https://doi.org/10.1007/s00125-003-1122-8.

Wells, L., Vosseller, K., & Hart, G. W. (2003). A role for N-acetylglucosamine as a nutrient sensor and mediator of insulin resistance. *Cellular and Molecular Life Sciences, 60*(2), 222–228.

Xu, Q., Niu, X., Wang, W., Yang, W., Du, Y., Gu, J., et al. (2017). Specific N-glycan alterations are coupled in EMT induced by different density cultivation of MCF 10A epithelial cells. *Glycoconjugate Journal, 34*(2), 219–227. https://doi.org/10.1007/s10719-016-9754-3.

Xu, Q., Qu, C., Wang, W., Gu, J., Du, Y., et al. (2017). Specific N-glycan alterations are coupled in epithelial-mesenchymal transition induced by EGF in GE11 epithelial cells. *Cell Biology International, 41*(2), 124–133. https://doi.org/10.1002/cbin.10707.

Ye, J., Wei, X., Shang, Y., Pan, Q., Yang, M., Tian, Y., et al. (2017). Core 3 mucin-type O-glycan restoration in colorectal cancer cells promotes MUC1/p53/miR-200c-dependent epithelial identity. *Oncogene, 36*(46), 6391–6407. https://doi.org/10.1038/onc.2017.241.

Yochum, Z. A., Cades, J., Mazzacurati, L., Neumann, N. M., Khetarpal, S. K., Chatterjee, S., et al. (2017). A first-in-class TWIST1 inhibitor with activity in oncogene-driven lung cancer. *Molecular Cancer Research, 15*(12), 1764–1776. https://doi.org/10.1158/1541-7786.MCR-17-0298.

Yochum, Z. A., Cades, J., Wang, H., Chatterjee, S., Simons, B. W., O'Brien, J. P., et al. (2019). Targeting the EMT transcription factor TWIST1 overcomes resistance to EGFR inhibitors in EGFR-mutant non-small-cell lung cancer. *Oncogene, 38*(5), 656–670. https://doi.org/10.1038/s41388-018-0482-y.

Zhang, W., Bouchard, G., Yu, A., Shafiq, M., Jamali, M., Shrager, J. B., et al. (2018). GFPT2-expressing cancer-associated fibroblasts mediate metabolic reprogramming in human lung adenocarcinoma. *Cancer Research, 78*(13), 3445–3457. https://doi.org/10.1158/0008-5472.CAN-17-2928.

Zhang, Y., Yu, X., Ichikawa, M., Lyons, J. J., Datta, S., Lamborn, I. T., et al. (2014). Autosomal recessive phosphoglucomutase 3 (PGM3) mutations link glycosylation defects to atopy, immune deficiency, autoimmunity, and neurocognitive impairment. *The Journal of Allergy and Clinical Immunology, 133*(5), 1400–1409. 1409.e1401–1405. https://doi.org/10.1016/j.jaci.2014.02.013.

Zheng, X., Carstens, J. L., Kim, J., Scheible, M., Kaye, J., Sugimoto, H., et al. (2015). Epithelial-to-mesenchymal transition is dispensable for metastasis but induces chemoresistance in pancreatic cancer. *Nature, 527*(7579), 525–530. https://doi.org/10.1038/nature16064.

Zhou, F., Su, J., Fu, L., Yang, Y., Zhang, L., Wang, L., et al. (2008). Unglycosylation at Asn-633 made extracellular domain of E-cadherin folded incorrectly and arrested in endoplasmic reticulum, then sequentially degraded by ERAD. *Glycoconjugate Journal, 25*(8), 727–740. https://doi.org/10.1007/s10719-008-9133-9.

CHAPTER THREE

The second genome: Effects of the mitochondrial genome on cancer progression

Adam D. Scheid, Thomas C. Beadnell, Danny R. Welch*
Department of Cancer Biology, The University of Kansas Medical Center, and The University of Kansas Cancer Center, Kansas City, KS, United States
*Corresponding author: e-mail address: dwelch@kumc.edu

Contents

1. Mitochondrial evolution and genetic variation — 65
2. Mitochondria and cancer — 67
   2.1 The Warburg effect — 67
   2.2 Reactive oxygen species — 69
   2.3 Oncometabolites — 71
   2.4 Mitochondrial morphology — 72
   2.5 mtDNA mutations and haplotype predispositions — 74
3. Studying direct mtDNA contributions to disease — 75
   3.1 Cybrids — 76
   3.2 Transmitochondrial mouse models — 78
   3.3 Conplastic mouse models — 80
   3.4 The MNX mouse model — 81
4. Concluding remarks and remaining questions — 90
Acknowledgments — 91
Conflict of interest — 91
References — 91

Abstract

The role of genetics in cancer has been recognized for centuries, but most studies elucidating genetic contributions to cancer have understandably focused on the nuclear genome. Mitochondrial contributions to cancer pathogenesis have been documented for decades, but how mitochondrial DNA (mtDNA) influences cancer progression and metastasis remains poorly understood. This lack of understanding stems from difficulty isolating the nuclear and mitochondrial genomes as experimental variables, which is critical for investigating direct mtDNA contributions to disease given extensive crosstalk exists between both genomes. Several in vitro and in vivo models have isolated mtDNA as an independent variable from the nuclear genome. This review compares and contrasts different models, their advantages and disadvantages for studying mtDNA contributions to cancer, focusing on the mitochondrial-nuclear exchange (MNX) mouse model and findings regarding tumor progression, metastasis, and other complex cancer-related phenotypes.

The vast majority of genetic studies in cancer focus on the nuclear genome, where numerous genetic and epigenetic alterations that contribute to tumorigenesis and metastasis have been characterized (Esquela-Kerscher & Slack, 2006; Goldberg et al., 2003; Lee et al., 1996; Lee & Welch, 1997; Phillips et al., 1996; Seraj, Samant, Verderame, & Welch, 2000; Steeg & Theodorescu, 2007; Weinstein & Joe, 2006; Welch et al., 1994). While nucleus-focused studies have provided critical insights into cancer initiation and progression, they, for the most part, have failed to consider the second genome harbored in eukaryotic cells: the mitochondrial genome. Otto Warburg's description of cancer cell aerobic glycolysis nearly a century ago set the stage for subsequent studies that have linked cancer and altered mitochondrial function (Brandon, Baldi, & Wallace, 2006; Chandra & Singh, 2011; Vyas, Zaganjor, & Haigis, 2016), but much has yet to be learned about mitochondrial contributions to cancer.

One of the challenging aspects in studying mitochondrial contributions to cancer is its inextricable link to the nuclear genome. At fewer than 17,000 base pairs, mitochondrial DNA (mtDNA) encodes only a fraction of the molecules required to carry out all the physiological functions in which the organelle is involved (Pagliarini et al., 2008; Taanman, 1999). Many of the other molecules necessary for mitochondrial function are encoded in nuclear DNA (nDNA), and accordingly oncogenic mutations in nDNA have significant impacts on mitochondrial biology (Nagarajan, Malvi, & Wajapeyee, 2016). Therefore, in order to study direct contributions of mtDNA on cancer and metastasis, mtDNA and nDNA must be isolated as separate experimental variables. To isolate mtDNA from nDNA contributions in vivo, we generated mitochondrial nuclear exchange (MNX) mice, in which mtDNA from various mouse strains can be combined with nDNA of other mouse strains (Kesterson et al., 2016). In this review, we detail how aberrant mitochondrial function contributes to tumor progression and metastasis, and how MNX mice have been utilized to clarify novel roles of mtDNA in cancer and metastasis as well as complex cancer-related phenotypes.

1. Mitochondrial evolution and genetic variation

Phylogenetic analyses suggest that mitochondria originate from a bacterium that developed an endosymbiotic relationship with the ancient unicellular host that phagocytosed it (Andersson et al., 1998; Ferla, Thrash, Giovannoni, & Patrick, 2013; Fitzpatrick, Creevey, & McInerney, 2006; Sassera et al., 2011; Wang & Wu, 2015; Yang, Oyaizu, Oyaizu, Olsen, & Woese, 1985). Although the identity of the precise bacterial ancestor remains controversial (Martijn, Vosseberg, Guy, Offre, & Ettema, 2018), the remains of unique bacterial traits, such as formylated proteins (Carp, 1982; Zhang et al., 2010), make clear the roots of mitochondria in bacterial ancestry. Extensive co-evolution between mitochondria and their hosts have resulted in an organelle that is central not only to its canonical role in metabolism and energy production but also to cell signaling, regulation of apoptosis, and many other critical cellular functions as well (Chandel, 2014; Martinou & Youle, 2011).

Another relic from the mitochondrial bacterial ancestor is the relatively small, circular mtDNA genome. The genome, which is composed of ~16,500 base pairs in humans and 16,300 base pairs in mice, encodes for 22 tRNAs, 2 rRNAs, and 13 proteins (Anderson et al., 1981). These proteins are all part of the electron transport chain (ETC) that resides within mitochondria, and include NADH dehydrogenase (ND)1, ND2, ND3, ND4, ND4L, ND5, and ND6 (complex I); cytochrome B (CYB, complex III); cytochrome c oxidase (CO)I, COII, and COIII (complex IV); and ATP synthase subunits 6 and 8 (ATP6 and ATP8, complex V) (Chomyn et al., 1986, 1985; Macreadie et al., 1983). Other components of the ETC, as well as machinery for mtDNA replication, transcription, and other critical mitochondrial functions are encoded in nDNA.

Counterintuitively, given the importance of the ETC for cellular energy and viability, the mutation rate of mtDNA is relatively high (Brown, George, & Wilson, 1979; Parsons et al., 1997). This high mutation rate, however, serves as a mechanism by which selective pressures can induce evolutionary adaptations. Accordingly, selection of mtDNA variants within ancient human populations enabled evolutionary adaptations to the various climates that those populations encountered while migrating to different regions of the planet. The selective pressure of climate on mtDNA evolution is evident in phylogenetic analysis of human mtDNA variants, where related variants, known as haplogroups, cluster according to geographic location (Balloux, Handley, Jombart, Liu, & Manica, 2009; Wallace, 2015). The first haplogroup, termed L0, originated in Africa, and the divergence of additional haplogroups began in Africa 130,000–170,000 years ago. Two haplogroups, M and N, diverged from the L3 haplogroup in Africa and went on to populate the rest of the world (Wallace, 2015). Divergence of mtDNA variants has been documented in wild and inbred laboratory mice as well, where introductions of the latter into relatively clean, controlled laboratory environments over the last century have likely played a major role mtDNA evolution (Goios, Pereira, Bogue, Macaulay, & Amorim, 2007).

The necessity of mtDNA adaptive evolution to climate is a logical one, considering that different climates are accompanied by different metabolic demands. Indeed, haplotype-defining mtDNA variants alter mitochondrial functions in ways that would aid in adaptation to new climates. For example, coupling efficiency, the efficiency with which the ETC generates proton gradients for ATP production by complex V, varies by haplotype, where more efficient haplotypes burn fewer calories per ATP generated and vice versa. Therefore, haplogroups in which more heat is generated per ATP are advantageous in cold climates, while haplotypes with better coupling efficiencies are more beneficial in warm climates (Kazuno et al., 2006; Wallace, 2015).

While high mtDNA mutation rates confer advantages for evolutionary adaptation, they also have the potential to increase the risk of incurring mutations that damage the ETC. This risk is largely mitigated by mtDNA copy number and inheritance during cell division. Eukaryotic cells can contain thousands of mitochondria per cell, and within each mitochondrion resides several copies of mtDNA (Fernandez-Vizarra, Enriquez, Perez-Martos, Montoya, & Fernandez-Silva, 2011; Robin & Wong, 1988; Shuster, Rubenstein, & Wallace, 1988). Therefore, unlike nDNA mutations that persist in all cell progeny, mtDNA mutations are limited to single organelles. This results in mixtures of normal and mutant mtDNA copies in each cell, a phenomenon collectively known as heteroplasmy.

Heteroplasmy and the nature of mtDNA inheritance limit the effects of both germline and somatic mtDNA mutations. mtDNA is exclusively inherited from maternal oocytes. During mitotic cell division, mitochondria and the mtDNA copies they harbor are distributed randomly (and unevenly) among daughter oocytes, and rounds of division and mtDNA replication lead to oocytes with various heteroplasmic ratios (Wallace & Chalkia, 2013). Oocytes enriched with mtDNA mutations that alter global mitochondrial function in the cell are selectively eliminated in the ovary, preventing dissemination of the mutations (Fan et al., 2008; Stewart et al., 2008). Similarly, mtDNA with somatic mutations are often outnumbered by normal mtDNA and are further diluted during cell division.

Just as mtDNA variations in organismal populations allow adaptations to new environments, they also allow cancer cells to adapt to dynamic microenvironments during tumorigenesis and metastasis (Brandon et al., 2006; Wallace, 2012, 2016). Metabolic adaptations are critical to tumor cell growth, as tumorigenesis can produce extreme microenvironmental alterations involving hypoxia and limited availability of other nutrients. Metastasis can induce even more extreme changes, where tumor cells must adapt to entirely new extracellular environments in order to colonize distant tissues. As discussed below, these adaptations to these dynamic microenvironments involve mutations in both mtDNA and nDNA.

2. Mitochondria and cancer

The importance of mitochondrial contributions to tumorigenesis and metastasis is underscored by recent observations that whole mitochondria are transferred from normal cells to tumor cells (Dong et al., 2017; Lu et al., 2017; Pasquier et al., 2013; Tan et al., 2015). Furthermore, Ishikawa and colleagues demonstrated that transferring mitochondria from tumor cells with high metastatic potential to tumor cells with low metastatic potential enhances the metastatic potential of the recipient cells and vice versa (Ishikawa et al., 2008). The interplay between oncogenic signaling and mitochondrial function is complex and, as described below, involves and extends beyond the role of mitochondria as the main energy producers of the cell.

2.1 The Warburg effect

Earliest indications of mitochondrial involvement in cancer came with Otto Warburg's discovery of aerobic glycolysis, commonly known as the Warburg effect, in cancer cells in the 1920s (Warburg, Wind, & Negelein, 1927). This discovery apparently contradicted Louis Pasteur's observation in 1861 that the presence of oxygen promotes more rapid cell division and inhibits fermentation in yeast, indicating that cancer cells displayed altered metabolism relative to healthy cells. Many groups have since described the Warburg effect in cancer cells, although it is not a universal trait among all tumor cell types (Dang, 2010; Herst & Berridge, 2007; Suganuma et al., 2010; Vander Heiden & DeBerardinis, 2017).

Warburg's explanation of aerobic glycolysis was predicated on the idea that mitochondrial oxidative phosphorylation (OXPHOS) was irreversibly defective in cancer cells, forcing the cells to perform glycolysis to produce ATP (Warburg, 1956a, 1956b). While impaired OXPHOS can underlie aerobic glycolysis (Lopez-Rios et al., 2007; Owens, Kulawiec, Desouki, Vanniarajan, & Singh, 2011), such defects are not necessarily irreversible. For example, Fantin et al. showed that knocking down lactate dehydrogenase A (LDH-A) stimulated ATP generation through OXPHOS in cancer cells that preferentially rely on aerobic glycolysis for energy production, demonstrating that neoplastic cells maintain the capacity to perform OXPHOS (Fantin, St-Pierre, & Leder, 2006). Furthermore, upregulated OXPHOS has been observed in multiple cancer types, further exemplifying that OXPHOS too can be utilized in oncogenic metabolism (Birkenmeier et al., 2016; Caro et al., 2012; Jones et al., 2016; Lagadinou et al., 2013; Viale et al., 2014; Whitaker-Menezes et al., 2011).

Many studies since Warburg's observations have provided mechanistic insights into aerobic glycolysis in cancer cells. Aerobic glycolysis provides a pathway through which cancer cells can continue to grow in hypoxic conditions. Hypoxic tumor microenvironments result from rapid tumor cell growth that can outpace vascularization (Chen et al., 2018; Semenza, 2013) and OXPHOS, due to reliance on oxygen reduction to produce ATP, is inefficient under hypoxic conditions. Although OXPHOS generates more ATP molecules per cycle ($n=36$) than does glycolysis ($n=2$), glycolysis produces ATP more rapidly than does OXPHOS, which is another aspect of glycolysis that benefits rapidly dividing cells (Pfeiffer, Schuster, & Bonhoeffer, 2001).

Aerobic glycolysis, like many other aspects of tumor cells, can be heterogeneous, enabling dynamic metabolic symbiosis within growing tumors. One example of metabolic symbiosis through heterogeneous aerobic glycolysis can occur between tumor cells in vascularized, oxygenated microenvironments and tumor cells in hypoxic environments (Corbet et al., 2018). Here, oxygenated tumor cells undergo OXPHOS and allow glucose from circulation to travel to tumor cells in hypoxic regions. In return, lactate from the glycolytic tumor cells travels back to the tumor cells undergoing OXPHOS, in which it can be converted to pyruvate by LDH-B and enter the mitochondria for OXPHOS and ATP generation (Bonuccelli et al., 2010; Griguer, Oliva, & Gillespie, 2005). Similarly, tumor cells can induce a glycolytic phenotype in surrounding fibroblasts and use the resulting lactate for OXPHOS in a phenomenon known as the reverse Warburg effect (Bonuccelli et al., 2010; Pavlides et al., 2009).

In addition to providing a metabolic advantage to tumor cells, aerobic glycolysis can modulate the microenvironment in ways that promote tumor progression and metastasis. For example, excess lactate production results in an acidic microenvironment, which has been shown to promote invasion and metastasis (Gatenby & Gawlinski, 1996). While tumor cells can accrue mutations to persist in such an acidic microenvironment, other cells that could limit tumor growth, such as immune cells, undergo apoptosis in low pH microenvironments (Park, Lyons, Ohtsubo, & Song, 1999). Similarly, high rates of glucose uptake for glycolysis competitively inhibit functionality of activated T cells that also undergo glycolysis, preventing elimination of tumor cells by the immune system (Cham, Driessens, O'Keefe, & Gajewski, 2008; Chang et al., 2013; Macintyre et al., 2014).

Multiple mutations and epigenetic alterations underlying the Warburg effect have been described. One such mechanism is hexokinase II expression in hepatomas, as opposed to hexokinase IV expression that occurs in normal hepatic cells. Hexokinase II has a lower Km than hexokinase IV, and can reside on the outer mitochondrial membrane (OMM) where it can use outgoing OXPHOS-generated ATP to phosphorylate glucose to glucose-6-phosphate and initiate glycolysis (Bustamante & Pedersen, 1977). Impairment of OXPHOS due to mutations in nDNA and mtDNA-encoded ETC components can drive aerobic glycolysis in cancer cells as well. Mutations in the mtDNA displacement (D)-loop, a major control locus for mtDNA replication and transcription (Taanman, 1999), downregulate transcription of mtDNA-encoded ETC components by lowering mtDNA copy number and transcription efficiency (Coskun, Beal, & Wallace, 2004; Lee et al., 2004). Similarly, mutations in mtDNA and nDNA genes encoding proteins involved in mtDNA maintenance and the ETC abrogate OXPHOS (Gasparre et al., 2007; Lopez-Rios et al., 2007; Owens et al., 2011; Singh, Ayyasamy, Owens, Koul, & Vujcic, 2009).

2.2 Reactive oxygen species

Mitochondria, in addition to NADPH oxidase, are the major source of reactive free radicals in cells. These include reactive oxygen species (ROS), a misnomer which also encompasses hydrogen peroxide as well as superoxide and hydroxide radicals, as well as reactive nitrogen species, such as nitric oxide and nitric dioxide radicals (Wiseman & Halliwell, 1996). ROS, in particular, are generated frequently in mitochondria as byproducts of

OXPHOS, where electron carriers of the ETC can transfer unpaired electrons in oxygen in the mitochondrial matrix. Mitochondria are equipped with a number of nDNA-encoded antioxidants, including superoxide dismutase, glutathione, thioredoxin, and peroxiredoxins, to protect mitochondrial DNA, proteins, and lipids from ROS-mediated oxidative damage (Hanschmann, Godoy, Berndt, Hudemann, & Lillig, 2013).

Increased ROS levels are frequently seen in cancer cells and, as with aerobic glycolysis, increases are often due to OXPHOS impairment. More specifically, mutations of ETC components that abrogate the flow of electrons through the ETC often induce increased levels of ROS (Mattiazzi et al., 2004). These mutations can impact the ability of ETC complexes to accept electrons, leaving the electrons on electron carriers and making them available for ROS generation (Ishii et al., 1998; Senoo-Matsuda et al., 2001).

ROS can augment tumor progression and metastasis through oxidative damage of macromolecules. mtDNA, given its proximity to ROS and lack of protection by histones, is particularly susceptible to ROS-mediated oxidative damage, and resulting mutations can contribute to increased ROS production and aerobic glycolysis as discussed above (Liemburg-Apers, Willems, Koopman, & Grefte, 2015; Lu, Sharma, & Bai, 2009). nDNA can sustain oxidative damage by ROS as well, which can result in mutations with the potential to further promote tumor progression and metastasis (Sallmyr et al., 2008).

In addition to its role as a genomic mutagen, ROS can also promote tumorigenesis by acting as mitogenic signaling molecules. More specifically, hydrogen peroxide is required for multiple signaling pathways that stimulate cell growth and division, including those involved in cytokine, insulin, and growth factor signaling as well as NF-κB signaling (Chandel, Trzyna, McClintock, & Schumacker, 2000; Krieger-Brauer & Kather, 1992; Lo & Cruz, 1995; Schreck, Rieber, & Baeuerle, 1991; Sundaresan, Yu, Ferrans, Irani, & Finkel, 1995). ROS-mediated signal augmentation is achieved through oxidation of thiol-containing cysteine residues within protein tyrosine phosphatase active sites, which promotes phosphorylation and activation of signaling molecules (Cunnick, Dorsey, Mei, & Wu, 1998; Denu & Tanner, 1998; Lee, Kwon, Kim, & Rhee, 1998). Depending on the target proteins, such alterations can provide mitogenic signals to promote tumor growth. Indeed, ROS-mediated oxidation of cysteine residues in proteins such as PTEN and Src provides mitogenic signals in cancer (Lee et al., 2002; Leslie et al., 2003).

While limited ROS concentrations can provide oncogenic signals through mutagenic and mitogenic means, excessive ROS concentrations and corresponding oxidative damage can induce cell death (Redza-Dutordoir & Averill-Bates, 2016). Therefore, tumor cells must maintain a balanced concentration of ROS to take advantage of their oncogenic properties without undergoing apoptosis. To achieve this balance, expression of antioxidant pathway molecules is often upregulated in tumor cells (DeNicola et al., 2011). This point is particularly important in metastasis, where treatment with antioxidants has actually been shown to cause increased metastatic efficiency, presumably due to attenuation of increased ROS levels associated with the metabolic demands of metastasis (Le Gal et al., 2015; Piskounova et al., 2015).

2.3 Oncometabolites

Just as nDNA-encoded components are instrumental for mtDNA maintenance and mitochondrial function, so too are mtDNA and mitochondrial functions critical for epigenetic nDNA organization. nDNA organization is in large part mediated by specific chemical modifications to deoxynucleotides as well as histone tails, where dynamic enzymatic modifications allow chromatin loosening and condensation based on the transcriptional requirements of the cell. These reactions require availability of substrates such as acetyl groups, which are in part provided by mitochondrial metabolism. For example, acetyltransferases can acquire acetyl groups from acetyl-coenzyme A (CoA), which is generated as a substrate for the tricarboxylic acid (TCA) cycle in mitochondria (Wellen et al., 2009).

In addition to participating in crosstalk with the nucleus in normal physiological states, mitochondrial metabolites function in oncogenic mitochondrial-nuclear crosstalk. The most well-characterized examples of this crosstalk involve molecules that structurally resemble α-ketoglutarate (α-KG), a metabolite in the TCA cycle. Aside from the TCA cycle α-KG serves as a substrate for dioxygenases, a superfamily of enzymes whose functions include epigenetic modifications and chromatin remodeling (Gerken et al., 2007). Mutations in isocitrate dehydrogenase, succinate dehydrogenase, and fumarate hydratase, which have been documented in several cancers (Alam et al., 2005; Baysal et al., 2000; Gimm, Armanios, Dziema, Neumann, & Eng, 2000; Parsons et al., 2008), result in excesses of metabolites that resemble α-KG (2-hydroxyglutarate, succinate, and fumarate, respectively). These metabolites can compete with α-KG for binding at dioxygenase active sites, inhibiting dioxygenase function. In some cancers, this results in epigenetic silencing of gene expression via hypermethylation at CpG islands (Noushmehr et al., 2010).

2.4 Mitochondrial morphology

Mitochondria, rather than existing as static organelles, exhibit dynamic morphologies that are dependent upon cellular status. Mitochondrial morphology is governed by two main processes: fusion and fission. Fusion is mediated by GTPases mitofusin 1 (Mfn1), Mfn2, and optic atrophy 1 (Opa1), where Mfn1 and 2 facilitate OMM fusion and Opa1 facilitates inner mitochondrial membrane (IMM) fusion to merge multiple mitochondria into a single, tubular organelle (Cipolat, Martins de Brito, Dal Zilio, & Scorrano, 2004; Santel & Fuller, 2001). Fission is mediated by dynamic related protein 1 (Drp1), a GTPase that binds receptors on the OMM and constricts the membrane to fragment a single mitochondrial network into multiple organelles (Smirnova, Shurland, Ryazantsev, & van der Bliek, 1998).

Mitochondrial structure has major impacts on mitochondrial function, and as such mitochondrial structure is dependent on the needs of the cell. Mitochondrial fusion is generally associated with OXPHOS, which is demonstrated by the observation that cells in nonfermentable conditions tend to display elongated mitochondrial networks (Egner, Jakobs, & Hell, 2002; Rossignol et al., 2004). These networks are thought to occur because fusion results in relatively large numbers of mtDNA copies in mitochondrial networks, enhancing production of mtDNA-encoded components of the ETC (Chen, Chomyn, & Chan, 2005; Chen et al., 2010). Oxidative stress also promotes mitochondrial fusion, perhaps as a mechanism to disperse ROS that are produced as byproducts of OXPHOS (Shutt, Geoffrion, Milne, & McBride, 2012). Conversely, mitochondrial fission is typically observed upon inhibition of OXPHOS, and fission decreases OXPHOS coupling efficiency (Wikstrom et al., 2014).

Aberrations in mitochondrial morphology have been reported in cancer cells. Skewing toward mitochondrial fission is often seen in cancer cells (Hagenbuchner, Kuznetsov, Obexer, & Ausserlechner, 2013; Inoue-Yamauchi & Oda, 2012; Kashatus et al., 2015; Rehman et al., 2012; Wan et al., 2014; Zhao et al., 2013), a logical observation given that OXPHOS is downregulated in many tumor cell types and fission is associated with OXPHOS inhibition (Wikstrom et al., 2014). While mitochondrial fission is not a universal trait of tumor cells (von Eyss et al., 2015), it can be critical to tumor progression and metastasis, as Drp1 inhibition and Mfn2 overexpression can impair tumor cell growth (Inoue-Yamauchi & Oda, 2012; Rehman et al., 2012) and upregulated Drp1 expression has been associated with a migratory phenotype in tumor cells (Ferreira-da-Silva et al., 2015).

The impacts of morphology on mitochondrial function extend beyond metabolism. Mitochondrial fission confers several other advantages to tumor cells. One advantage is resistance to apoptosis, a hallmark of tumor cells. Mitochondria play integral roles in initiation of apoptosis. In response to various signals, pro-apoptotic Bcl-2 family members Bax and Bak oligomerize to induce OMM permeabilization (MOMP) which, in turn, results in cytochrome c release into the cytoplasm, leading to activation of proteolytic caspases that execute apoptotic pathways. In addition to upregulated expression of anti-apoptotic members of the Bcl-2 family (Strasser, Harris, Bath, & Cory, 1990; Tsujimoto, Finger, Yunis, Nowell, & Croce, 1984), increased mitochondrial fission functions as a mechanism by which tumor cells escape apoptosis, where hyperfragmentation inhibits Bax interactions with the OMM (Renault et al., 2015).

Another mechanism by which skewed mitochondrial fission can be oncogenic is through induction of increased mitophagy. Mitophagy is the process by which mitochondria are cleared from the cell, often due to damage and dysfunction. Dysfunctional mitochondria tend to have depolarized membranes due to inability of the ETC to generate proton gradients across the IMM, and this depolarization allows Pink1 kinase to accumulate at the OMM and phosphorylate mitochondrial surface proteins (Matsuda et al., 2010; Narendra et al., 2010). This results in recruitment and activation of Parkin (Okatsu et al., 2015), an E3 ligase that ubiquitinates mitochondrial surface proteins, resulting in degradation of the proteins and targeting of the mitochondria for autophagic membranes (Chan et al., 2011; Sarraf et al., 2013). Fission can potentiate mitophagy simply by decreasing mitochondrial size, and increased mitophagy can help established tumors adapt to new environments and engender therapeutic resistance (Hu et al., 2012). The relationship between mitophagy and cancer is complex, however, as mitophagy can also inhibit tumor growth (Lee et al., 2012; Tay et al., 2010), indicating that mitophagy can be pro- or anti-tumorigenic depending on the dynamic needs of the tumor.

Mitochondrial spatial dynamics are critical to tumor cell progression and metastasis as well (Attanasio et al., 2011; Caino et al., 2016; Desai, Bhatia, Toner, & Irimia, 2013). One of the most well-documented examples was demonstrated by Altieri and colleagues, who showed that Akt reactivation in tumor cells treated with a phosphoinositide 3-kinase (PI3K) inhibitor results in translocation of mitochondria to the cortical cytoskeleton (Caino et al., 2015). This translocation resulted in increased lamellipodia dynamics and focal adhesion complex turnover, which combined to augment tumor cell migration and invasion (Caino et al., 2015). Notably, Mfn1 and OXPHOS inhibition ameliorated both mitochondrial translocation as well as increased tumor cell migration and invasion, suggesting respiration is critical for both organellar translocation as well as the resulting phenotypes (Caino et al., 2015). Mitochondrial translocation is also dependent on syntaphilin (SNPH), where ubiquitinated SNPH inhibits mitochondrial dynamics and translocation (Caino et al., 2016; Seo et al., 2018).

2.5 mtDNA mutations and haplotype predispositions

Interestingly, the mtDNA mutations that contribute to the altered, oncogenic functionality of mitochondria described above can occur throughout the mitochondrial genome. Thus, a somatic mutation in a region defining a particular haplotype can "convert" the sequence to another haplotype, thereby possibly confusing interpretation of haplotype-dependent susceptibility (Brandon et al., 2006, 2005). While incidence of the latter observation may be overestimated due to sequencing errors, as parallel sequencing of normal and tumor tissue from the same individual is not performed in most studies, its occurrence has been definitively demonstrated (Parrella et al., 2001). Large insertions or deletions that give rise to changes in conserved amino acids can have drastic impacts on mtDNA and mitochondrial function, e.g., truncation of ETC components that may be important for early tumorigenesis. Mutagenic conversions that match non-self mtDNA haplotype sequences, however, may be important for metabolic adaptations to dynamic tumor microenvironments, much like divergent mtDNA haplotypes were important for adaptations to new climates in ancient peoples.

Understanding how mtDNA haplotype variants contribute to tumorigenesis and cancer progression is important for two main reasons. One reason is that individuals with particular mtDNA haplotypes have increased predispositions for developing certain cancers relative to individuals with other mtDNA haplotypes (Brinker et al., 2017; Bussard & Siracusa, 2017; Feeley et al., 2015). Indeed, adaptive advantages that mtDNA variants confer can also resemble oncogenic mitochondrial function discussed above (Ross et al., 2001; van der Walt et al., 2003). Better understandings of these predispositions can enable more effective cancer screening and prevention. The second reason why a better understanding of how mtDNA variants contribute to cancer is critical is because it can precipitate development of therapeutic interventions that block the ability of tumors to adapt to changing microenvironments, which may halt tumor growth and prevent therapeutic resistance.

Since nDNA-encoded components are instrumental to mtDNA maintenance and mitochondrial function, querying direct contributions of mtDNA to cancer requires separating nDNA and mtDNA as isolated variables. As outlined below, there are several ways these variables can be isolated in vitro in mouse and man. However, corresponding in vivo studies in humans present major ethical barriers. These studies can be performed in vivo in mouse models, but traditional backcrossing on female genetic backgrounds to obtain conplastic mice with nDNA from one mouse strain and mtDNA from another can introduce confounding recombinations in nDNA. To address this issue, we generated MNX mice by exchanging embryonic pronuclei among mouse strains (Fetterman et al., 2013; Kesterson et al., 2016), and the resulting model has enabled and will continue to enable novel insights on direct mtDNA contributions to cancer and other complex phenotypes that can interact with the disease.

3. Studying direct mtDNA contributions to disease

The unique characteristics of mtDNA relative to its nuclear counterpart make mtDNA genetic manipulation and corresponding functional studies challenging. For example, the presence of hundreds to thousands of mtDNA molecules within single cells makes alteration of all mtDNA copies extremely difficult. Therefore, traditional approaches to mtDNA engineering would likely result in heteroplasmy, and random distribution of altered mtDNA among daughter cells would produce variable heteroplasmic ratios among manipulated cells, confounding effects of the alteration.

Although methods for mtDNA genetic engineering are being explored (Gammage, Rorbach, Vincent, Rebar, & Minczuk, 2014; Hashimoto et al., 2015; Jo et al., 2015; Patananan, Wu, Chiou, & Teitell, 2016; Reddy et al., 2015; Trifunovic et al., 2004), existing in vitro and in vivo models for studying mtDNA contributions to disease are predicated on transferring mtDNA molecules with polymorphisms/mutations of interest to cells with the same nDNA as cells containing control mtDNA. These models have enabled studies elucidating mtDNA contributions to mitochondrial physiology and disease such as those discussed above. Current efforts, including our generation of the MNX mouse model, are focused on improving these models, with the goals of facilitating a better understanding of mtDNA biology and harnessing gained knowledge to develop improved therapies for diseases to which mtDNA contributes.

3.1 Cybrids

One of the first major innovations that enabled interrogation of direct mtDNA impacts on phenotypic traits was the development of cytoplasmic hybrid, or cybrid, cells. Cybrids are generated by fusing nucleated cells with enucleated cells (often platelets), resulting in transfer of cytoplasmic contents of the enucleated cell, including mtDNA, to the nucleated cell. This is as opposed to hybrid cells, which are the products of fusing nucleated cells.

The first cybrid cells that were produced were generated using nucleated cells that were replete with mtDNA, resulting in heteroplasmic mixtures of enucleated cell and nucleated cell mtDNA (Bunn, Wallace, & Eisenstadt, 1974). While cybrids have proven useful (Wallace, Bunn, & Eisenstadt, 1975), more direct experiments querying how mtDNA sequence variants influenced phenotypic traits required generation of cybrids homoplasmic for those mtDNA sequences. This was accomplished by the development of rho-null (ρ^0) cells, which are nucleated cells lacking a mitochondrial genome.

The feasibility of mtDNA depletion was demonstrated by the observation that yeast naturally reduced mtDNA copy numbers under conditions that favored glycolysis over OXPHOS (Wilkins, Carl, & Swerdlow, 2014). Subsequently, several methods by which mtDNA can be artificially depleted from cells have been developed. The first was incubation with ethidium bromide (EtBr), a positively charged compound that can enter negatively charged mitochondrial matrices and intercalate into mtDNA to inhibit its replication. After several decades of work the first human ρ^0 cell line was derived from the 143B osteosarcoma cell line in 1989 (King & Attardi, 1989). Since that time additional human ρ^0 cell lines, and comparatively less toxic methods to generate them, have been produced as reviewed elsewhere (Wilkins et al., 2014).

Rather than outright destroying mtDNA, agents used to generate ρ^0 cell lines inhibit its replication. Therefore, as cell division progresses in the presence of the agent, mtDNA is diluted among daughter cells until no mtDNA remains. Mitochondrial remnants can still be found in ρ^0 cells, but their functionality is severely altered (Swerdlow et al., 1996). Most notably, since they lack the mtDNA-encoded components of the ETC they are incapable of OXPHOS and rely solely on glycolysis for energy production. In addition, upon complete depletion of mtDNA, human ρ^0 cells become auxotrophic for uridine and pyruvate due to lack of ETC function and glycolytic oxidation-reduction requirements, respectively (Gregoire, Morais, Quilliam, & Gravel, 1984; King & Attardi, 1989, 1996).

The use of ρ^0 cell lines has been critical for clarifying many aspects of mitochondrial biology, from establishing potential links between mtDNA sequence variants and phenotypes to uncovering how mitochondrial dysfunction contributes to various diseases. For example, cybrid studies using mtDNA from patients with Leber's hereditary optic neuropathy (LHON), which is associated with mutations in mtDNA ND genes encoding complex I components, revealed connections between mutant mtDNA and deficiencies in oxygen consumption and complex I function (Baracca et al., 2005; Jun, Trounce, Brown, Shoffner, & Wallace, 1996). Cybrid studies have been useful for investigating mtDNA contributions to non-classical mitochondrial diseases as well. Such is the case with Parkinson's disease, for which cybrid studies have demonstrated that mtDNA is at least partially responsible for decreases in complex I function that have been documented in the disease (Esteves et al., 2008, 2010; Gu, Cooper, Taanman, & Schapira, 1998; Swerdlow et al., 1998).

Cybrid studies have also been instrumental for demonstrating how nDNA and mtDNA backgrounds (as well as crosstalk between nDNA and mtDNA) influence phenotypic manifestations of mitochondrial diseases. For example, cybrids have been used to show that variable OXPHOS kinetics among mtDNA haplotypes contribute to discrepant susceptibilities of those haplotypes to LHON (Pello et al., 2008). In addition, the potency with which the mitochondrial encephalopathy, lactic acidosis, and stroke-like episodes syndrome (MELAS)-associated A3243G inhibits cytochrome oxidase function is dependent on the nDNA background of the ρ^0 cell line used to study it (Dunbar, Moonie, Jacobs, & Holt, 1995). Furthermore, Picard et al. elucidated a molecular mechanism underlying the heterogeneity with which MELAS presents clinically by using cybrid cell lines to show that various heteroplasmic ratios of the A3243G mutation resulted in differential transcriptomic profiles from both nDNA and mtDNA (Picard et al., 2014).

While cybrids have proven to be useful tools for studying mitochondrial biology and mtDNA contributions to disease, they do have several important limitations. In addition to the caveat of all in vitro models, namely that observations may not recapitulate what occurs in normal physiological settings, only limited ρ^0 cell lines have been created and performing experiments in the most relevant cell type may be challenging (Wilkins et al., 2014). Most ρ^0 cell lines that have been created are derived from tumor cells as well, which may further confound results due to nDNA instability and corresponding potential expression level alterations of nDNA-encoded mitochondrial components (as well as other nDNA-encoded genes). Finally, although it is not an issue limited to cybrid models, it is extremely difficult to associate single nucleotide polymorphisms (SNP) with phenotypic observations. mtDNA tends to harbor variability in wide ranges of nucleotides among individuals, and mtDNA SNPs are extremely difficult to isolate given the challenges associated with mtDNA engineering. This lack of ability to isolate SNPs is confounding because, as demonstrated using cybrid models themselves (Pello et al., 2008), the functional consequences of SNP are influenced by the mtDNA haplotypes in which they exist. Microheteroplasmy, which is the presence of relatively small proportions (1–2%) of mutant mtDNA molecules that can be difficult to detect, can influence phenotypic traits as well (Smigrodzki & Khan, 2005). These limitations notwithstanding, cybrids still serve as valuable tools for understanding mtDNA biology.

3.2 Transmitochondrial mouse models

To enable examination of mtDNA contributions to disease in vivo, mtDNA transgenic mice were created. Transmitochondrial mice, often referred to as "mito-mice," were first generated using microinjection of mitochondria containing mtDNA of interest into zygotes followed by implantation into nDNA-matched females (Irwin, Johnson, & Pinkert, 1999; Pinkert, Irwin, Johnson, & Moffatt, 1997). Subsequent techniques harnessed advancements that were used in the generation of cybrids and ρ^0 cells. More specifically, ρ^0 embryonic stem (ES) cells were made using rhodamine 6G, fused with cytoplasts containing the mtDNA of interest, and either injected into zygotes or co-cultured with blastocysts, after which the chimeric germ cells were implanted into nDNA-matched females (Inoue et al., 2000; Irwin et al., 1999; Marchington, Barlow, & Poulton, 1999; Sligh et al., 2000). Although first generation progeny from each of these methods contained heteroplasmic ratios of transgenic and endogenous mtDNA, homoplasmic transmitochondrial mice were attainable by breeding transmitochondrial females with nDNA-matched males (Sligh et al., 2000).

Given the ability to transfer any desired mtDNA into ρ^0 mouse ES cells, transmitochondrial mice provide an ideal model to study physiological effects of pathogenic mtDNA. Several homoplasmic transmitochondrial mouse models have been successfully created, the first being mice containing mtDNA conferring chloramphenicol resistance (CAP^R) (Levy, Waymire, Kim, MacGregor, & Wallace, 1999; Marchington et al., 1999). These mice were derived by fusing CAP^R cytoplasts harboring the T2433C mtDNA 16S rRNA mutation with mouse ES cells in culture (Blanc, Wright, Bibb, Wallace, & Clayton, 1981; Bunn et al., 1974; Levy et al., 1999), and the resulting progeny displayed striking pathogenic phenotypes including growth retardation and in utero or perinatal lethality (Sligh et al., 2000). Similarly, transmitochondrial mice containing mtDNA with a 4696 base pair mtDNA deletion have been generated, and these mice display mitochondrial dysfunction in multiple tissues prior to mortality, often due to renal failure (Inoue et al., 2000).

Another way in which the transmitochondrial mouse model has been used to study effects of mtDNA variants is through generation of xenomitochondrial mice. Several xenocybrids, including human ρ^0 cells containing primate mtDNA or mouse ρ^0 cells containing *Rattus norvegicus* mtDNA, have been created (Dey, Barrientos, & Moraes, 2000; Kenyon & Moraes, 1997; McKenzie & Trounce, 2000; Yamaoka et al., 2000). Each displayed significant OXPHOS defects, likely owing to incompatibility of mismatched nDNA and mtDNA-encoded ETC components (Barrientos, Kenyon, & Moraes, 1998; Dey et al., 2000; McKenzie & Trounce, 2000; Yamaoka et al., 2000). McKenzie et al. combined the xenocybrid and transmitochondrial mouse models to generate xenomitochondrial mice, in which mtDNA from *Mus spretus* and *Mus dunni* mice was transferred to ρ^0 *Mus musculus domesticus* ES cells (McKenzie, Trounce, Cassar, & Pinkert, 2004). OXPHOS was largely unaltered in the xenomitochondrial mice, but increased glycolysis indicated that the xenomitochondrial cells were more glycolytic than their wild-type counterparts (McKenzie et al., 2004).

Despite the great potential of transmitochondrial mouse models, their utility has been limited by one major factor: a paucity of mutant mtDNA sequences to study. Few natural murine pathogenic mtDNA mutations are known, and current inability to engineer pathogenic mtDNA mutations hampers the ability to create pathogenic mtDNA. This may soon change, as random mtDNA mutagenesis has been achieved through homozygous knock-in of a proofreading-deficient mtDNA polymerase γ subunit, Polgα (Trifunovic et al., 2004), and directed mtDNA mutagenesis using mitochondria-targeted restriction endonucleases, transcription activator-like effector nucleases (mitoTALEN), zinc finger nucleases (mtZFN), and clustered regularly interspaced short palindromic repeats (CRISPR)/Cas9 hold promise (Gammage et al., 2014; Hashimoto et al., 2015; Jo et al., 2015; Reddy et al., 2015).

3.3 Conplastic mouse models

The most traditional genetic way in which to study the physiological consequences of mtDNA variation is the conplastic mouse. Leveraging exclusive maternal inheritance of mtDNA, conplastic mice are generated by breeding female mice with mtDNA of interest with male mice harboring the nDNA background of interest. Female F_1 progeny, containing the desired mtDNA and equal contributions of nDNA from maternal and paternal sources, are then backcrossed with male mice containing the original paternal nDNA. This cross is performed for at least 10 generations, resulting in conplastic mice with mtDNA from one inbred mouse strain and 99.9% nDNA from another (Markel et al., 1997).

Rather than studying the effects of known pathogenic mtDNA in vivo as with transmitochondrial mice, conplastic mouse models examine how natural variations in mtDNA impact mitochondrial function and downstream phenotypes. Such a model is highly relevant to the human condition, where divergent mtDNA haplotypes differentially influence mitochondrial function and can consequentially predispose individuals to disease (Canter, Kallianpur, Parl, & Millikan, 2005; Liu et al., 2003). Ibrahim and colleagues, in an exhaustive phylogenetic study, demonstrated that mtDNA divergence has also occurred in inbred laboratory mouse strains, where 50 of 52 strains tested contained mtDNA that diverged from a single *Mus musculus domesticus* female ancestor (Yu et al., 2009). These mtDNA variants, although not directly analogous to human mtDNA haplotypes, provide a platform with which conplastic models can be used to understand how mtDNA variants influence mitochondrial function and relevant phenotypes in vivo.

Conplastic mice have been used to demonstrate previously unappreciated mtDNA contributions to complex phenotypic traits as well as diseases. For example, using conplastic mice with mtDNA from the FVB/NJ and NZB/BlnJ strains on the C57BL/6J nDNA background alongside wild-type C57BL/6 mice, Hirose et al. demonstrated that a SNP in ATP8, the only SNP between FVB/NJ and C57BL/6J mtDNA, altered microbial profiles present in the intestine (Hirose et al., 2017). While previous associations between human mtDNA haplotypes and intestinal microbiota had been established

(Ma et al., 2014), the conplastic model allowed for a direct association without the confounding factor of disparate nDNA backgrounds. The relevance of conplastic models to the human condition has also been demonstrated by the observation that various mouse mtDNA backgrounds impart differential susceptibility to experimental autoimmune encephalomyelitis (EAE) (Yu et al., 2009). EAE is a model that recapitulates the etiology and symptoms of multiple sclerosis (MS) (Constantinescu, Farooqi, O'Brien, & Gran, 2011), and consistent with the finding that mtDNA influences EAE in conplastic mice human mtDNA variants have been associated with MS susceptibility (Yu et al., 2008).

Conplastic mice also provide a useful model for understanding the consequences of mtDNA replacement, which is critical given recent interest in using mitochondrial replacement to prevent mitochondrial diseases in humans (Craven et al., 2010; Ma et al., 2015). Accordingly, a comprehensive study by Latorre-Pellicer et al. demonstrated how complex the ramifications of mtDNA replacement are, as a multitude of phenotypes, most notably those involving metabolism and aging, were altered upon replacement of C57BL/6JOlaHsd mtDNA with that of NZB/OlaHsd on the C57BL/6JOlaHsd nDNA background (Latorre-Pellicer et al., 2016). Such changes are indicative of the complexity involved in nuclear-mitochondrial crosstalk. Changes in sperm motility among conplastic strains independent of ATP production or polymorphism load (number of SNP per mtDNA) exemplify this complexity as well (Tourmente et al., 2017).

Although conplastic mouse models largely eliminate disparate nDNA backgrounds as confounding variables in mtDNA studies, numerous backcrosses necessary to derive the mice are accompanied by a higher probability for introduction of nDNA recombination that muddles comparisons between conplastic mice and their wild-type counterparts. The backcrosses are also relatively time-consuming, which can make studies using conplastic strains long and expensive.

3.4 The MNX mouse model

To circumvent the issues associated with the approaches above, we generated MNX mice. Unlike transmitochondrial and conplastic mouse models, which require cybrids and extensive backcrossing, respectively, MNX mice with unaltered nDNA (i.e., not exposed to mutagens) and homoplasmic mtDNA (also not mutagen-exposed) are generated relatively rapidly via pronuclear transfer.

To begin, super-ovulated dams with the desired mtDNA are mated with nDNA-matched males. The resulting embryos are harvested from the oviducts, pronuclei are isolated from embryos of each strain using micropipettes, and the extracted pronuclei are transferred to embryos of the other strain. The embryos are then implanted into pseudopregnant females and brought to term, after which F_1 females are bred with nDNA-matched males to propagate the MNX strain. Successful mtDNA transfer and homoplasmy are evaluated using restriction fragment length polymorphisms (Kesterson et al., 2016).

In addition to the relatively rapid time in which MNX mice can be generated, the major advantage that the MNX model offers over transmitochondrial and conplastic models is reduced potential for introducing confounding variables. For example, many of the transmitochondrial mouse models that have been created remain heteroplasmic, which makes direct attributions of observed phenotypes to a particular mtDNA sequence impossible. Off-target effects from using rhodamine 6G to generate cybrids for transmitochondrial mouse production may influence observed phenotypes as well. In conplastic models extensive backcrossing involves nDNA recombination that could alter nuclear-mitochondrial crosstalk, which, in addition to mtDNA variants, could influence observed phenotypes. Thus, we believe that the MNX model has significant advantages for probing novel aspects of mtDNA and mitochondrial biology.

The impetus for generating the MNX mouse model was rooted in a study by Kent Hunter and colleagues that aimed to determine the impact of genetics on tumor latency and metastatic efficiency (Lifsted et al., 1998). They crossed females from 27 different inbred mouse strains to male FVB/N mice harboring a transgene encoding the oncogenic polyomavirus middle T antigen (PyMT) under the control of the mouse mammary tumor virus (MMTV) promoter (FVB/N-TgN(MMTV-PyMT)). Expression of the oncogenic transgene results in spontaneous mammary gland tumors that readily metastasize to the lungs, providing an excellent model for studying both tumor growth and dissemination (Guy, Cardiff, & Muller, 1992). Crossing the inbred strains with the FVB/N-TgN(MMTV-PyMT) mouse revealed striking ranges in latency to tumor outgrowth as well as metastatic burden among the strains, demonstrating the influence of genetic factors on tumor progression and metastasis (Lifsted et al., 1998).

Since establishing a role for mouse genetic backgrounds in tumor latency and metastatic efficiency, Hunter and colleagues employed backcrossing and global genetic screens to identify nDNA-encoded metastasis modifiers,

(Faraji et al., 2014; Ha, Long, Cai, Shu, & Hunter, 2016). These modifiers exemplify the complexities of the biological processes underlying metastasis, as they encode proteins whose functions range from immune cell interactions to circadian rhythm maintenance (Faraji et al., 2014, 2012; Ha et al., 2016). Importantly, the results have also uncovered homologous metastasis modifiers in human breast cancer, demonstrating the relevance of the model to human disease (Herschkowitz et al., 2007; Hsieh, Look, Sieuwerts, Foekens, & Hunter, 2009; Pfefferle et al., 2013).

Given maternal transmission of mtDNA and the Hunter group's original experimental design, crossing females from various strains to male FVB/N-TgN(MMTV-PyMT) mice, we hypothesized that mtDNA was contributing to the observed phenotypes as well. To test this hypothesis, we crossed male FVB/N-TgN(MMTV-PyMT) mice with wild-type female FVB/NJ mice (FF mice, Table 1) as well as two female MNX strains containing FVB/NJ nDNA: one with mtDNA from C57BL/6J mice and another with mtDNA from BALB/cJ mice (FC and FB mice, respectively, Table 1) (Feeley et al., 2015). Our results were strikingly similar to those reported by Lifsted et al., where FC mice displayed longer tumor latency and lower metastatic burden and FB mice had shorter tumor latency and more metastatic burden than FF mice (Table 1) (Feeley et al., 2015; Lifsted et al., 1998). This observation, for the first time in a spontaneous tumor model, showed that mtDNA does indeed influence tumor progression and metastasis.

It is important to emphasize that mtDNA polymorphisms are likely metastasis *modifiers* rather than drivers *per se*. mtDNA encodes quantitative trait loci (QTL) that combine with both nuclear and mitochondrially encoded genes to regulate complex diseases like cancer and disease severity (Cookson, Liang, Abecasis, Moffatt, & Lathrop, 2009; Hunter, Amin, Deasy, Ha, & Wakefield, 2018; Zhang et al., 2018). QTL are a group of alleles that influence a particular phenotype or trait (Abiola et al., 2003). Measurable trait differences can be influenced by the combined interactions of multiple different polymorphisms present within the genome, as well as environmental factors. This is exemplified by observations that nDNA backgrounds influence the same mtDNA in different ways (Dunbar et al., 1995), and that mtDNA polymorphisms do not tend to exhibit strong maternal inheritance patterns. Instead, nuclear-mitochondrial crosstalk as well as individual responses to changing environmental factors are more likely influenced by mitochondrial polymorphisms.

While pulmonary metastatic burden differed between FF, FB and FC strains, the number of metastases was not significantly different among

Table 1 MNX mice and corresponding mtDNA-directed phenotypes.

nDNA	mtDNA	Abbreviation	Tumor latency[a,b]	Metastatic Size[a,b]	Metastatic number[a,b]	Epigenetic changes[c]	Cardiac volume overload[d]
FVB/NJ	BALB/cJ	FB	PyMT: ↓ Her2: ↑	PyMT: ↑ Her2: ↑	PyMT: NS Her2: ↓	Yes	ND
FVB/NJ	C57BL/6J	FC	PyMT: ↑ Her2: ↑	PyMT: ↓ Her2: ↑	PyMT: NS Her2: ↓	Yes	ND
C57BL/6J	C3H/HeN	CH	ND	ND	ND	Yes	Resistant
C3H/HeN	C57BL/6J	HC	ND	ND	ND	Yes	Sensitive

[a]Feeley et al. (2015).
[b]Brinker et al. (2017).
[c]Vivian et al. (2017).
[d]Fetterman et al. (2013).
Phenotypes are relative to wild-type strains with matching nDNA. *ND*, not done; *NS*, not significant.

the strains (Table 1) (Feeley et al., 2015; Lifsted et al., 1998). This observation suggested that the mtDNA polymorphisms among the strains (Fig. 1) affected tumor cell growth at the metastatic site, potentially through differences in ability to adapt to dynamic metabolic demands at the new site. Accordingly, tumor cells from FC mice displayed significantly higher basal oxygen consumption rates (OCR) than did tumor cells from FB or FF mice, but significantly lower reserve capacity than those observed in the other

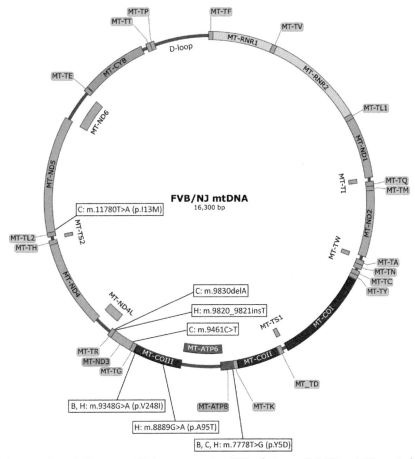

Fig. 1 mtDNA SNP among MNX mouse strains. SNP relative to FVB/NJ mtDNA are indicated in boxes, with initial letters indicating mouse strains as in Table 1 and amino acid changes indicated in parentheses where applicable. mtDNA map was adapted from SnapGene® Viewer 4.1.6, and mtDNA sequences were obtained from NCBI Nucleotide at the following GenBank accessions: B: AJ512208.1, C: DQ106412.1, F: EF108338.1, H: EF108335.1.

strains (Feeley et al., 2015). Since reserve capacity has been linked with ability to adapt to metabolic stress (Dranka, Hill, & Darley-Usmar, 2010), such as that associated with metastasis and colonization of distant tissue sites, diminished reserve capacity and consequential difficulty in adapting to metabolic stress may have been responsible for the decreased metastatic burden in FC mice relative to the other strains.

Metabolic alterations owing to mtDNA polymorphisms among the MNX strains may have contributed to the noted differences in tumor latency as well. Tumor cells from FC mice had higher ratios of OCR to extracellular acidification rates (ECAR) than did the other strains, suggesting an increased dependence on OXPHOS as opposed to glycolysis in tumor cells from those mice relative to the other strains (Feeley et al., 2015). A more respiratory metabolic profile could hinder adaptation to rapid tumor cell growth and associated hypoxia, which could slow tumor outgrowth. Tumor cells from FC mice were capable of glycolysis, however, even exhibiting the highest ECAR among the strains (Feeley et al., 2015), but transitioning from relying more on OXPHOS than glycolysis to a more glycolytic phenotype may slow metabolic adaptation and tumor outgrowth.

Another interesting observation that has been made using the MNX mouse model is that mtDNA influences on tumor progression and metastatic efficiency are oncogenic driver-dependent. This was demonstrated by crossing female FF, FB, and FC mice with male FVB/N-Tg(MMTVneu) mice over-expressing wild-type Her2/neu oncogene under the control of the MMTV promoter (Brinker et al., 2017). While the results between FF and FC mice were like those seen in the PyMT model, Her2 FB mice had unexpectedly longer tumor latency and lower metastatic burden than FF mice (Table 1). In both MNX crosses, numbers of pulmonary metastases were significantly lower relative to FF mice (Table 1) (Brinker et al., 2017). Importantly, the Her2 study also definitively demonstrated that observed changes in tumor progression were directly attributable to mtDNA and not another cytoplasmic element among the MNX mice, as crossing male FB and FC mice with female FVB/N-Tg(MMTVneu) mice resulted in female offspring whose tumor latency did not significantly differ (Brinker et al., 2017).

That mtDNA impacts tumor progression and metastasis in an oncogenic driver-dependent manner is novel but not necessarily surprising. PyMT and Her2 expression induce two distinct breast cancer subtypes (Pfefferle et al., 2013). Therefore, it stands to reason that nuclear-mitochondrial crosstalk may be altered in each tumor type. Indeed, oncogene-specific signaling between the nucleus and mitochondria has been demonstrated. For example, c-Myc and mammalian target of rapamycin (mTOR) drive transformation of many tumor cell types, and while both modulate expression of nDNA-encoded mitochondrial components the targets and modulatory mechanisms differ between the signaling molecules. c-Myc increases mitochondrial biogenesis in part by upregulation transcription of hundreds of nDNA-encoded mitochondrial genes (Li et al., 2005), while mTOR increases mitochondrial biogenesis through specific transcriptional upregulation of peroxisome proliferator-activated receptor gamma coactivator-1 alpha (PGC-1α) and Yin Yang 1 (YY1) and resulting increased translation of mitochondrial mRNAs through inhibition of 4E binding proteins (4E-BPs) (Cunningham et al., 2007). Oncogenes can also manipulate mitochondrial biology by controlling mitochondrial dynamics (Caino et al., 2016; Kashatus et al., 2015; Serasinghe et al., 2015). This too exhibits oncogene specificity, as oncogenic K-Ras promotes mitochondrial fission through Drp1 phosphorylation (Kashatus et al., 2015; Serasinghe et al., 2015), while c-Myc promotes mitochondrial fusion (von Eyss et al., 2015).

It is important to consider that in the MNX mouse breast cancer studies detailed above mtDNA substitutions were made not only in the tumor cells but also in the rest of the cells in the mice as well. This raises the possibility that mtDNA effects on the observed phenotypes were products of both tumor cell intrinsic as well as tumor cell extrinsic mechanisms. Addressing this possibility, unpublished work from our laboratory demonstrates that transplantation of tumor cells with canonical mtDNA into nDNA-matched (i.e., syngeneic) MNX mice with various mtDNA backgrounds alters metastatic efficiency (A. E. Brinker & D. R. Welch, in preparation), indicating that mtDNA also influences metastasis through tumor cell extrinsic mechanisms. Preliminarily, ROS neutralization normalized differences in metastatic efficiency in these studies, suggesting mtDNA-mediated ROS production from non-tumor cells impacts metastatic efficiency (A. E. Brinker & D. R. Welch, in preparation).

One way in which mtDNA could regulate complex, polygenic phenomena such as tumor progression and metastasis in tumor cell intrinsic and extrinsic manners is through modulation nDNA epigenetics and subsequent gene expression (Bellizzi, D'Aquila, Giordano, Montesanto, & Passarino, 2012; Smiraglia, Kulawiec, Bistulfi, Gupta, & Singh, 2008; Xie et al., 2007). To test this possibility for the first time in vivo, we performed integrative global genomic analyses to look for variations in nDNA methylation and gene expression in brain tissue from MNX mice derived from the

BALB/cJ, C57BL/6J, FVB/NJ, and C3H/HeN backgrounds (Table 1) (Vivian et al., 2017) as well as histone marks (J. T. McGuire & D. R. Welch, in preparation). Three-way comparisons including wild-type "parental" strains and MNX mice revealed specific contributions of mtDNA to variations in both nDNA methylation as well as gene expression, and in some cases increased methylation status corresponded with downregulated gene expression as expected. While these observations were made using brain tissue, which fails to exactly recapitulate epigenetic landscapes most relevant for the mammary tumors used for our studies and their corresponding sites of metastasis (lung and lymph node), the results solidify epigenetic modulation as a mechanism by which mtDNA impacts tumor progression and metastasis.

Rather than global alterations, changes in DNA methylation among the MNX strains were limited to specific loci. The mechanism for such specificity is unclear given mtDNA-mediated epigenetic regulation is thought to occur through contributions to global metabolite pools that can serve as substrates for epigenetic-modifying enzymes (Lozoya et al., 2018; Wellen et al., 2009). One potential mechanism may be mtDNA-directed localization of DNA methyltransferases (DNMTs) that could differ among mtDNA backgrounds. Another intriguing possibility that oncogene-directed localization of DNMTs, in concert with variable contributions to global epigenetic substrates by different mtDNA backgrounds, serves as a mechanism by which mtDNA impacts on tumor progression and metastasis are oncogene dependent.

Modulating the immune system could exert tumor cell intrinsic (e.g., antigen presentation or checkpoint regulator expression) or extrinsic effects on tumor progression and metastasis (e.g., immune cell differentiation or effector function). The immune system plays a complex, dichotomous role in cancer. Recognition of inflammatory danger signals and mutated host peptides can instigate immune-mediated tumor cell clearance, but induction of immune tolerance by tumor cells and other immune cells promotes immune evasion and tumor growth (Mittal, Gubin, Schreiber, & Smyth, 2014). Anti-tumor immunotherapy, a booming area of cancer therapy that has provided some promising results (Couzin-Frankel, 2013), leverages both aspects of this relationship (Brahmer et al., 2012; Hodi et al., 2010; Porter, Levine, Kalos, Bagg, & June, 2011; Topalian et al., 2012).

Immune cell activation and differentiation impose dynamic metabolic demands, and differential responsiveness to those demands by polymorphic mtDNA could alter overall immune responses to inflammatory stimuli

(Buck, Sowell, Kaech, & Pearce, 2017). The dynamic metabolic requirements required for immune cell function are exemplified by effector and memory T cells, both of which are critical for anti-tumor immunity. Memory T cells, which persist after primary immune responses to surveil for reemergence of cognate foreign or mutated antigens, tend to display mitochondrial fusion and rely on OXPHOS to fuel cell functions. Effector T cells, which are short-lived, inflammatory cells that respond to acute immune insult, tend to display mitochondrial fission and rely on aerobic glycolysis to fuel cell functions (Buck et al., 2016). Because many tumor cell types also perform aerobic glycolysis, glucose consumption by tumor cells represents a tumor cell intrinsic mechanism by which mtDNA could modulate immunity, where glucose depletion by tumor cells can inhibit effector T cell function (Cham et al., 2008; Chang et al., 2013; Macintyre et al., 2014).

Several studies using MNX mice have implicated that mtDNA (in) directly influences the immune system. Fetterman et al. used MNX mice with C57BL/6J and C3H/HeN genetic backgrounds to test mtDNA impacts on atherosclerosis and cardiac valve regurgitation (Fetterman et al., 2013), both pathologies which have been associated with macrophage infiltration (Brands et al., 2013). ROS production and membrane polarization in cardiomyocytes or mice harboring C57BL/6J mtDNA were more susceptible to both phenotypes than cells/mice containing C3H/HeN mtDNA (Table 1) (Fetterman et al., 2013). Similarly, Betancourt et al. demonstrated that mtDNA influences aspects of atherogenic diet-induced non-alcoholic fatty liver disease (NAFLD), as MNX mice with C57BL/6 and C3H/HeN mtDNA displayed intermediate phenotypes relative to their wild-type counterparts (Betancourt et al., 2014). Aberrant macrophage function has also been associated with NAFLD (Chatterjee et al., 2013; Huang et al., 2010; Lanthier et al., 2011; Rivera et al., 2007). Furthermore, initial studies have revealed mtDNA-dependent differences in immune cell differentiation and proportion (T. C. Beadnell & D. R. Welch, unpublished data).

Finally, mtDNA may also influence tumor progression and metastasis in a tumor extrinsic manner by altering commensal microbiota. mtDNA haplotypes have been associated with commensal microbial profiles (Hirose et al., 2017; Ma et al., 2014), and both cancer and cancer therapy responsiveness have been linked to microbiota (Farrell et al., 2012; Routy et al., 2018; Vetizou et al., 2015; Viaud et al., 2013; Wang et al., 2012). The relationship between mtDNA and commensal microbes is multi-faceted and complex. One facet is host metabolism, which affects many aspects of host physiology and can therefore impact commensal microbes in a multitude of ways. Bacterial products, by processes such as quorum sensing, can affect host mitochondria as well, which could dictate commensal microbe colonization in a mtDNA-dependent manner (Tao et al., 2016). Additionally, given the cyclical nature of immune system interactions with microbiota (Maynard, Elson, Hatton, & Weaver, 2012), mtDNA-mediated immune system dynamics may alter microbiota and, conversely, mtDNA-mediated microbial colonization could modulate immune system maturation and/or evolution. Indeed, ongoing work indicates that intestinal microbiota differs among MNX mouse strains (S. J. Manley & D. R. Welch, in preparation), suggesting that mtDNA influences commensal microbe colonization in the MNX mouse model.

4. Concluding remarks and remaining questions

Querying direct, physiologically relevant mtDNA contributions to phenotypes and diseases is extremely difficult in humans due to confounding nDNA heterogeneity, but cybrids and mouse models have proven indispensable for studying such contributions. The MNX mouse model is the first in which other confounding variables such as mutagens, heteroplasmy, and nDNA recombination have been eliminated, making it quite useful for in vivo modeling of mtDNA biology. Accordingly, MNX mice have been utilized to demonstrate novel mtDNA impacts on tumor initiation, tumor outgrowth, metastasis, and other cancer-relevant phenotypes including epigenetic, immune, and commensal microbial profiles (Table 1).

While the MNX mouse model has enabled many novel insights into mtDNA biology, its utilization has led to as many, if not more, questions about how such a relatively small genome can impact such complex, multigenic phenotypes. Of them all, perhaps an overarching question is the most intriguing: what mtDNA-regulated signals mediate influence over complex phenotypes? Work to this point has implicated ROS and metabolic substrates as potential mediators of metastatic potential and epigenetic landscapes, respectively. Are there other signals? What are they? Recently, it was discovered that mtDNA encodes small non-coding RNA (sncRNA) derived from coding or tRNA genes (Barrey et al., 2011; Ro et al., 2013; Sripada et al., 2012). Some of these sncRNA are associated with complex phenotypes such as cancer (Magee, Telonis, Loher, Londin, & Rigoutsos, 2018; Telonis et al., 2015; Telonis & Rigoutsos, 2018). Could polymorphic sncRNA differentially affect mtDNA contributions to complex phenotypes? MNX mice provide an ideal model with which to investigate these and many other questions.

In addition to shedding light on novel aspects of mtDNA biology, results from MNX mouse studies carry the potential to inform mtDNA-related disease prevention and therapeutic strategies in the future. In particular, since that the MNX mouse model more closely recapitulates *pre-existing* [emphasis added] mtDNA variants as opposed to somatic mtDNA mutants, MNX mice can be used to understand how mtDNA variation impacts basal mitochondrial function and how variable mitochondrial function confers disease susceptibility as a QTL. Exploration into these topics will be highly relevant to understanding how human mtDNA haplotypes impart disease susceptibility. The outcome could be valuable for individualized screening and/or therapeutic strategies. Opportunities like these may soon be made reality given increasing appreciation for, and study of, mtDNA as a genomic contributor to disease.

Acknowledgments

We are grateful for generous support from: Susan G. Komen for the Cure (SAC110037) and the National Foundation for Cancer Research. Additional funding support was provided by U.S. Army Medical Research Defense Command, W81XWH-18-1-0450 (T.C.B.); National Cancer Institute P30-CA168524 (D.R.W.) and National Institutes of Health GM103418 (T.C.B. and D.R.W.). We apologize to any authors whose work was omitted due to space limitations.

Conflict of interest

The authors declare no conflicts of interest.

References

Abiola, O., Angel, J. M., Avner, P., Bachmanov, A. A., Belknap, J. K., Bennett, B., et al. (2003). The nature and identification of quantitative trait loci: A community's view. *Nature Reviews. Genetics*, *4*(11), 911–916.

Alam, N. A., Barclay, E., Rowan, A. J., Tyrer, J. P., Calonje, E., Manek, S., et al. (2005). Clinical features of multiple cutaneous and uterine leiomyomatosis: An underdiagnosed tumor syndrome. *Archives of Dermatology*, *141*(2), 199–206.

Anderson, S., Bankier, A. T., Barrell, B. G., de Bruijn, M. H., Coulson, A. R., Drouin, J., et al. (1981). Sequence and organization of the human mitochondrial genome. *Nature*, *290*(5806), 457–465.

Andersson, S. G., Zomorodipour, A., Andersson, J. O., Sicheritz-Ponten, T., Alsmark, U. C., Podowski, R. M., et al. (1998). The genome sequence of Rickettsia prowazekii and the origin of mitochondria. *Nature*, *396*(6707), 133–140.

Attanasio, F., Caldieri, G., Giacchetti, G., van Horssen, R., Wieringa, B., & Buccione, R. (2011). Novel invadopodia components revealed by differential proteomic analysis. *European Journal of Cell Biology, 90*(2–3), 115–127.

Balloux, F., Handley, L. J., Jombart, T., Liu, H., & Manica, A. (2009). Climate shaped the worldwide distribution of human mitochondrial DNA sequence variation. *Proceedings of the Biological Sciences, 276*(1672), 3447–3455.

Baracca, A., Solaini, G., Sgarbi, G., Lenaz, G., Baruzzi, A., Schapira, A. H., et al. (2005). Severe impairment of complex I-driven adenosine triphosphate synthesis in leber hereditary optic neuropathy cybrids. *Archives of Neurology, 62*(5), 730–736.

Barrey, E., Saint-Auret, G., Bonnamy, B., Damas, D., Boyer, O., & Gidrol, X. (2011). Pre-microRNA and mature microRNA in human mitochondria. *PLoS One, 6*(5), e20220.

Barrientos, A., Kenyon, L., & Moraes, C. T. (1998). Human xenomitochondrial cybrids. Cellular models of mitochondrial complex I deficiency. *The Journal of Biological Chemistry, 273*(23), 14210–14217.

Baysal, B. E., Ferrell, R. E., Willett-Brozick, J. E., Lawrence, E. C., Myssiorek, D., Bosch, A., et al. (2000). Mutations in SDHD, a mitochondrial complex II gene, in hereditary paraganglioma. *Science, 287*(5454), 848–851.

Bellizzi, D., D'Aquila, P., Giordano, M., Montesanto, A., & Passarino, G. (2012). Global DNA methylation levels are modulated by mitochondrial DNA variants. *Epigenomics, 4*(1), 17–27.

Betancourt, A. M., King, A. L., Fetterman, J. L., Millender-Swain, T., Finley, R. D., Oliva, C. R., et al. (2014). Mitochondrial-nuclear genome interactions in non-alcoholic fatty liver disease in mice. *Biochemical Journal, 461*(2), 223–232.

Birkenmeier, K., Drose, S., Wittig, I., Winkelmann, R., Kafer, V., Doring, C., et al. (2016). Hodgkin and Reed-Sternberg cells of classical Hodgkin lymphoma are highly dependent on oxidative phosphorylation. *International Journal of Cancer, 138*(9), 2231–2246.

Blanc, H., Wright, C. T., Bibb, M. J., Wallace, D. C., & Clayton, D. A. (1981). Mitochondrial DNA of chloramphenicol-resistant mouse cells contains a single nucleotide change in the region encoding the 3′ end of the large ribosomal RNA. *Proceedings of the National Academy of Sciences of the United States of America, 78*(6), 3789–3793.

Bonuccelli, G., Tsirigos, A., Whitaker-Menezes, D., Pavlides, S., Pestell, R. G., Chiavarina, B., et al. (2010). Ketones and lactate "fuel" tumor growth and metastasis: Evidence that epithelial cancer cells use oxidative mitochondrial metabolism. *Cell Cycle, 9*(17), 3506–3514.

Brahmer, J. R., Tykodi, S. S., Chow, L. Q., Hwu, W. J., Topalian, S. L., Hwu, P., et al. (2012). Safety and activity of anti-PD-L1 antibody in patients with advanced cancer. *New England Journal of Medicine, 366*(26), 2455–2465.

Brandon, M., Baldi, P., & Wallace, D. C. (2006). Mitochondrial mutations in cancer. *Oncogene, 25*(34), 4647–4662.

Brandon, M. C., Lott, M. T., Nguyen, K. C., Spolim, S., Navathe, S. B., Baldi, P., et al. (2005). MITOMAP: A human mitochondrial genome database—2004 update. *Nucleic Acids Research, 33*(Database issue), D611–D613.

Brands, M., Roelants, J., de Krijger, R., Bogers, A., Reuser, A., van der Ploeg, A., et al. (2013). Macrophage involvement in mitral valve pathology in mucopolysaccharidosis type VI (Maroteaux-Lamy syndrome). *American Journal of Medical Genetics. Part A, 161A*(10), 2550–2553.

Brinker, A. E., Vivian, C. J., Koestler, D. C., Tsue, T. T., Jensen, R. A., & Welch, D. R. (2017). Mitochondrial haplotype alters mammary cancer tumorigenicity and metastasis in an oncogenic driver-dependent manner. *Cancer Research, 77*(24), 6941–6949.

Brown, W. M., George, M., Jr., & Wilson, A. C. (1979). Rapid evolution of animal mitochondrial DNA. *Proceedings of the National Academy of Sciences of the United States of America, 76*(4), 1967–1971.

Buck, M. D., O'Sullivan, D., Klein Geltink, R. I., Curtis, J. D., Chang, C. H., Sanin, D. E., et al. (2016). Mitochondrial dynamics controls T cell fate through metabolic programming. *Cell, 166*(1), 63–76.

Buck, M. D., Sowell, R. T., Kaech, S. M., & Pearce, E. L. (2017). Metabolic instruction of immunity. *Cell, 169*(4), 570–586.

Bunn, C. L., Wallace, D. C., & Eisenstadt, J. M. (1974). Cytoplasmic inheritance of chloramphenicol resistance in mouse tissue culture cells. *Proceedings of the National Academy of Sciences of the United States of America, 71*(5), 1681–1685.

Bussard, K. M., & Siracusa, L. D. (2017). Understanding mitochondrial polymorphisms in cancer. *Cancer Research, 77*(22), 6051–6059.

Bustamante, E., & Pedersen, P. L. (1977). High aerobic glycolysis of rat hepatoma cells in culture: Role of mitochondrial hexokinase. *Proceedings of the National Academy of Sciences of the United States of America, 74*(9), 3735–3739.

Caino, M. C., Ghosh, J. C., Chae, Y. C., Vaira, V., Rivadeneira, D. B., Faversani, A., et al. (2015). PI3K therapy reprograms mitochondrial trafficking to fuel tumor cell invasion. *Proceedings of the National Academy of Sciences of the United States of America, 112*(28), 8638–8643.

Caino, M. C., Seo, J. H., Aguinaldo, A., Wait, E., Bryant, K. G., Kossenkov, A. V., et al. (2016). A neuronal network of mitochondrial dynamics regulates metastasis. *Nature Communications, 7*, 13730.

Canter, J. A., Kallianpur, A. R., Parl, F. F., & Millikan, R. C. (2005). Mitochondrial DNA G10398A polymorphism and invasive breast cancer in African-American women. *Cancer Research, 65*(17), 8028–8033.

Caro, P., Kishan, A. U., Norberg, E., Stanley, I. A., Chapuy, B., Ficarro, S. B., et al. (2012). Metabolic signatures uncover distinct targets in molecular subsets of diffuse large B cell lymphoma. *Cancer Cell, 22*(4), 547–560.

Carp, H. (1982). Mitochondrial N-formylmethionyl proteins as chemoattractants for neutrophils. *The Journal of Experimental Medicine, 155*(1), 264–275.

Cham, C. M., Driessens, G., O'Keefe, J. P., & Gajewski, T. F. (2008). Glucose deprivation inhibits multiple key gene expression events and effector functions in CD8+ T cells. *European Journal of Immunology, 38*(9), 2438–2450.

Chan, N. C., Salazar, A. M., Pham, A. H., Sweredoski, M. J., Kolawa, N. J., Graham, R. L., et al. (2011). Broad activation of the ubiquitin-proteasome system by Parkin is critical for mitophagy. *Human Molecular Genetics, 20*(9), 1726–1737.

Chandel, N. S. (2014). Mitochondria as signaling organelles. *BMC Biology, 12*, 34.

Chandel, N. S., Trzyna, W. C., McClintock, D. S., & Schumacker, P. T. (2000). Role of oxidants in NF-kappa B activation and TNF-alpha gene transcription induced by hypoxia and endotoxin. *Journal of Immunology, 165*(2), 1013–1021.

Chandra, D., & Singh, K. K. (2011). Genetic insights into OXPHOS defect and its role in cancer. *Biochimica et Biophysica Acta, 1807*(6), 620–625.

Chang, C. H., Curtis, J. D., Maggi, L. B., Jr., Faubert, B., Villarino, A. V., O'Sullivan, D., et al. (2013). Posttranscriptional control of T cell effector function by aerobic glycolysis. *Cell, 153*(6), 1239–1251.

Chatterjee, S., Ganini, D., Tokar, E. J., Kumar, A., Das, S., Corbett, J., et al. (2013). Leptin is key to peroxynitrite-mediated oxidative stress and Kupffer cell activation in experimental non-alcoholic steatohepatitis. *Journal of Hepatology, 58*(4), 778–784.

Chen, H., Chomyn, A., & Chan, D. C. (2005). Disruption of fusion results in mitochondrial heterogeneity and dysfunction. *Journal of Biological Chemistry, 280*(28), 26185–26192.

Chen, A., Sceneay, J., Godde, N., Kinwel, T., Ham, S., Thompson, E. W., et al. (2018). Intermittent hypoxia induces a metastatic phenotype in breast cancer. *Oncogene, 37*(31), 4214–4225.

Chen, H., Vermulst, M., Wang, Y. E., Chomyn, A., Prolla, T. A., McCaffery, J. M., et al. (2010). Mitochondrial fusion is required for mtDNA stability in skeletal muscle and tolerance of mtDNA mutations. *Cell*, *141*(2), 280–289.

Chomyn, A., Cleeter, M. W., Ragan, C. I., Riley, M., Doolittle, R. F., & Attardi, G. (1986). URF6, last unidentified reading frame of human mtDNA, codes for an NADH dehydrogenase subunit. *Science*, *234*(4776), 614–618.

Chomyn, A., Mariottini, P., Cleeter, M. W., Ragan, C. I., Matsuno-Yagi, A., Hatefi, Y., et al. (1985). Six unidentified reading frames of human mitochondrial DNA encode components of the respiratory-chain NADH dehydrogenase. *Nature*, *314*(6012), 592–597.

Cipolat, S., Martins de Brito, O., Dal Zilio, B., & Scorrano, L. (2004). OPA1 requires mitofusin 1 to promote mitochondrial fusion. *Proceedings of the National Academy of Sciences of the United States of America*, *101*(45), 15927–15932.

Constantinescu, C. S., Farooqi, N., O'Brien, K., & Gran, B. (2011). Experimental autoimmune encephalomyelitis (EAE) as a model for multiple sclerosis (MS). *British Journal of Pharmacology*, *164*(4), 1079–1106.

Cookson, W., Liang, L., Abecasis, G., Moffatt, M., & Lathrop, M. (2009). Mapping complex disease traits with global gene expression. *Nature Reviews. Genetics*, *10*(3), 184–194.

Corbet, C., Bastien, E., Draoui, N., Doix, B., Mignion, L., Jordan, B. F., et al. (2018). Interruption of lactate uptake by inhibiting mitochondrial pyruvate transport unravels direct antitumor and radiosensitizing effects. *Nature Communications*, *9*(1), 1208.

Coskun, P. E., Beal, M. F., & Wallace, D. C. (2004). Alzheimer's brains harbor somatic mtDNA control-region mutations that suppress mitochondrial transcription and replication. *Proceedings of the National Academy of Sciences of the United States of America*, *101*(29), 10726–10731.

Couzin-Frankel, J. (2013). Breakthrough of the year 2013. Cancer immunotherapy. *Science*, *342*(6165), 1432–1433.

Craven, L., Tuppen, H. A., Greggains, G. D., Harbottle, S. J., Murphy, J. L., Cree, L. M., et al. (2010). Pronuclear transfer in human embryos to prevent transmission of mitochondrial DNA disease. *Nature*, *465*(7294), 82–85.

Cunnick, J. M., Dorsey, J. F., Mei, L., & Wu, J. (1998). Reversible regulation of SHP-1 tyrosine phosphatase activity by oxidation. *Biochemistry and Molecular Biology International*, *45*(5), 887–894.

Cunningham, J. T., Rodgers, J. T., Arlow, D. H., Vazquez, F., Mootha, V. K., & Puigserver, P. (2007). mTOR controls mitochondrial oxidative function through a YY1-PGC-1alpha transcriptional complex. *Nature*, *450*(7170), 736–740.

Dang, C. V. (2010). Rethinking the Warburg effect with Myc micromanaging glutamine metabolism. *Cancer Research*, *70*(3), 859–862.

DeNicola, G. M., Karreth, F. A., Humpton, T. J., Gopinathan, A., Wei, C., Frese, K., et al. (2011). Oncogene-induced Nrf2 transcription promotes ROS detoxification and tumorigenesis. *Nature*, *475*(7354), 106–109.

Denu, J. M., & Tanner, K. G. (1998). Specific and reversible inactivation of protein tyrosine phosphatases by hydrogen peroxide: Evidence for a sulfenic acid intermediate and implications for redox regulation. *Biochemistry*, *37*(16), 5633–5642.

Desai, S. P., Bhatia, S. N., Toner, M., & Irimia, D. (2013). Mitochondrial localization and the persistent migration of epithelial cancer cells. *Biophysical Journal*, *104*(9), 2077–2088.

Dey, R., Barrientos, A., & Moraes, C. T. (2000). Functional constraints of nuclear-mitochondrial DNA interactions in xenomitochondrial rodent cell lines. *The Journal of Biological Chemistry*, *275*(40), 31520–31527.

Dong, L. F., Kovarova, J., Bajzikova, M., Bezawork-Geleta, A., Svec, D., Endaya, B., et al. (2017). Horizontal transfer of whole mitochondria restores tumorigenic potential in mitochondrial DNA-deficient cancer cells. *eLife*, *6*, e22187.

Dranka, B. P., Hill, B. G., & Darley-Usmar, V. M. (2010). Mitochondrial reserve capacity in endothelial cells: The impact of nitric oxide and reactive oxygen species. *Free Radical Biology and Medicine*, *48*(7), 905–914.

Dunbar, D. R., Moonie, P. A., Jacobs, H. T., & Holt, I. J. (1995). Different cellular backgrounds confer a marked advantage to either mutant or wild-type mitochondrial genomes. *Proceedings of the National Academy of Sciences of the United States of America*, *92*(14), 6562–6566.

Egner, A., Jakobs, S., & Hell, S. W. (2002). Fast 100-nm resolution three-dimensional microscope reveals structural plasticity of mitochondria in live yeast. *Proceedings of the National Academy of Sciences of the United States of America*, *99*(6), 3370–3375.

Esquela-Kerscher, A., & Slack, F. J. (2006). Oncomirs—MicroRNAs with a role in cancer. *Nature Reviews. Cancer*, *6*(4), 259–269.

Esteves, A. R., Domingues, A. F., Ferreira, I. L., Januario, C., Swerdlow, R. H., Oliveira, C. R., et al. (2008). Mitochondrial function in Parkinson's disease cybrids containing an nt2 neuron-like nuclear background. *Mitochondrion*, *8*(3), 219–228.

Esteves, A. R., Lu, J., Rodova, M., Onyango, I., Lezi, E., Dubinsky, R., et al. (2010). Mitochondrial respiration and respiration-associated proteins in cell lines created through Parkinson's subject mitochondrial transfer. *Journal of Neurochemistry*, *113*(3), 674–682.

Fan, W., Waymire, K. G., Narula, N., Li, P., Rocher, C., Coskun, P. E., et al. (2008). A mouse model of mitochondrial disease reveals germline selection against severe mtDNA mutations. *Science*, *319*(5865), 958–962.

Fantin, V. R., St-Pierre, J., & Leder, P. (2006). Attenuation of LDH-A expression uncovers a link between glycolysis, mitochondrial physiology, and tumor maintenance. *Cancer Cell*, *9*(6), 425–434.

Faraji, F., Hu, Y., Wu, G., Goldberger, N. E., Walker, R. C., Zhang, J., et al. (2014). An integrated systems genetics screen reveals the transcriptional structure of inherited predisposition to metastatic disease. *Genome Research*, *24*(2), 227–240.

Faraji, F., Pang, Y., Walker, R. C., Nieves Borges, R., Yang, L., & Hunter, K. W. (2012). Cadm1 is a metastasis susceptibility gene that suppresses metastasis by modifying tumor interaction with the cell-mediated immunity. *PLoS Genetics*, *8*(9), e1002926.

Farrell, J. J., Zhang, L., Zhou, H., Chia, D., Elashoff, D., Akin, D., et al. (2012). Variations of oral microbiota are associated with pancreatic diseases including pancreatic cancer. *Gut*, *61*(4), 582–588.

Feeley, K. P., Bray, A. W., Westbrook, D. G., Johnson, L. W., Kesterson, R. A., Ballinger, S. W., et al. (2015). Mitochondrial genetics regulate breast cancer tumorigenicity and metastatic potential. *Cancer Research*, *75*(20), 4429–4436.

Ferla, M. P., Thrash, J. C., Giovannoni, S. J., & Patrick, W. M. (2013). New rRNA gene-based phylogenies of the Alphaproteobacteria provide perspective on major groups, mitochondrial ancestry and phylogenetic instability. *PLoS One*, *8*(12), e83383.

Fernandez-Vizarra, E., Enriquez, J. A., Perez-Martos, A., Montoya, J., & Fernandez-Silva, P. (2011). Tissue-specific differences in mitochondrial activity and biogenesis. *Mitochondrion*, *11*(1), 207–213.

Ferreira-da-Silva, A., Valacca, C., Rios, E., Populo, H., Soares, P., Sobrinho-Simoes, M., et al. (2015). Mitochondrial dynamics protein Drp1 is overexpressed in oncocytic thyroid tumors and regulates cancer cell migration. *PLoS One*, *10*(3), e0122308.

Fetterman, J. L., Zelickson, B. R., Johnson, L. W., Moellering, D. R., Westbrook, D. G., Pompilius, M., et al. (2013). Mitochondrial genetic background modulates bioenergetics and susceptibility to acute cardiac volume overload. *Biochemical Journal*, *455*(2), 157–167.

Fitzpatrick, D. A., Creevey, C. J., & McInerney, J. O. (2006). Genome phylogenies indicate a meaningful alpha-proteobacterial phylogeny and support a grouping of the mitochondria with the Rickettsiales. *Molecular Biology and Evolution*, *23*(1), 74–85.

Gammage, P. A., Rorbach, J., Vincent, A. I., Rebar, E. J., & Minczuk, M. (2014). Mitochondrially targeted ZFNs for selective degradation of pathogenic mitochondrial genomes bearing large-scale deletions or point mutations. *EMBO Molecular Medicine*, *6*(4), 458–466.

Gasparre, G., Porcelli, A. M., Bonora, E., Pennisi, L. F., Toller, M., Iommarini, L., et al. (2007). Disruptive mitochondrial DNA mutations in complex I subunits are markers of oncocytic phenotype in thyroid tumors. *Proceedings of the National Academy of Sciences of the United States of America*, *104*(21), 9001–9006.

Gatenby, R. A., & Gawlinski, E. T. (1996). A reaction-diffusion model of cancer invasion. *Cancer Research*, *56*(24), 5745–5753.

Gerken, T., Girard, C. A., Tung, Y. C., Webby, C. J., Saudek, V., Hewitson, K. S., et al. (2007). The obesity-associated FTO gene encodes a 2-oxoglutarate-dependent nucleic acid demethylase. *Science*, *318*(5855), 1469–1472.

Gimm, O., Armanios, M., Dziema, H., Neumann, H. P., & Eng, C. (2000). Somatic and occult germ-line mutations in *SDHD*, a mitochondrial complex II gene, in nonfamilial pheochromocytoma. *Cancer Research*, *60*(24), 6822–6825.

Goios, A., Pereira, L., Bogue, M., Macaulay, V., & Amorim, A. (2007). mtDNA phylogeny and evolution of laboratory mouse strains. *Genome Research*, *17*(3), 293–298.

Goldberg, S. F., Miele, M. E., Hatta, N., Takata, M., Paquette-Straub, C., Freedman, L. P., et al. (2003). Melanoma metastasis suppression by chromosome 6: Evidence for a pathway regulated by CRSP3 and TXNIP. *Cancer Research*, *63*(2), 432–440.

Gregoire, M., Morais, R., Quilliam, M. A., & Gravel, D. (1984). On auxotrophy for pyrimidines of respiration-deficient chick embryo cells. *European Journal of Biochemistry*, *142*(1), 49–55.

Griguer, C. E., Oliva, C. R., & Gillespie, G. Y. (2005). Glucose metabolism heterogeneity in human and mouse malignant glioma cell lines. *Journal of Neuro-Oncology*, *74*(2), 123–133.

Gu, M., Cooper, J. M., Taanman, J. W., & Schapira, A. H. (1998). Mitochondrial DNA transmission of the mitochondrial defect in Parkinson's disease. *Annals of Neurology*, *44*(2), 177–186.

Guy, C. T., Cardiff, R. D., & Muller, W. J. (1992). Induction of mammary tumors by expression of polyomavirus middle T oncogene: A transgenic mouse model for metastatic disease. *Molecular and Cellular Biology*, *12*(3), 954–961.

Ha, N. H., Long, J., Cai, Q., Shu, X. O., & Hunter, K. W. (2016). The circadian rhythm gene Arntl2 is a metastasis susceptibility gene for estrogen receptor-negative breast cancer. *PLoS Genetics*, *12*(9), e1006267.

Hagenbuchner, J., Kuznetsov, A. V., Obexer, P., & Ausserlechner, M. J. (2013). BIRC5/Survivin enhances aerobic glycolysis and drug resistance by altered regulation of the mitochondrial fusion/fission machinery. *Oncogene*, *32*(40), 4748–4757.

Hanschmann, E. M., Godoy, J. R., Berndt, C., Hudemann, C., & Lillig, C. H. (2013). Thioredoxins, glutaredoxins, and peroxiredoxins—molecular mechanisms and health significance: From cofactors to antioxidants to redox signaling. *Antioxidants and Redox Signaling*, *19*(13), 1539–1605.

Hashimoto, M., Bacman, S. R., Peralta, S., Falk, M. J., Chomyn, A., Chan, D. C., et al. (2015). MitoTALEN: A general approach to reduce mutant mtDNA loads and restore oxidative phosphorylation function in mitochondrial diseases. *Molecular Therapy*, *23*(10), 1592–1599.

Herschkowitz, J. I., Simin, K., Weigman, V. J., Mikaelian, I., Usary, J., Hu, Z., et al. (2007). Identification of conserved gene expression features between murine mammary carcinoma models and human breast tumors. *Genome Biology*, *8*(5), R76.

Herst, P. M., & Berridge, M. V. (2007). Cell surface oxygen consumption: A major contributor to cellular oxygen consumption in glycolytic cancer cell lines. *Biochimica et Biophysica Acta*, *1767*(2), 170–177.

Hirose, M., Kunstner, A., Schilf, P., Sunderhauf, A., Rupp, J., Johren, O., et al. (2017). Mitochondrial gene polymorphism is associated with gut microbial communities in mice. *Scientific Reports, 7*(1), 15293.

Hodi, F. S., O'Day, S. J., McDermott, D. F., Weber, R. W., Sosman, J. A., Haanen, J. B., et al. (2010). Improved survival with ipilimumab in patients with metastatic melanoma. *New England Journal of Medicine, 363*(8), 711–723.

Hsieh, S. M., Look, M. P., Sieuwerts, A. M., Foekens, J. A., & Hunter, K. W. (2009). Distinct inherited metastasis susceptibility exists for different breast cancer subtypes: A prognosis study. *Breast Cancer Research, 11*(5), R75.

Hu, Y. L., DeLay, M., Jahangiri, A., Molinaro, A. M., Rose, S. D., Carbonell, W. S., et al. (2012). Hypoxia-induced autophagy promotes tumor cell survival and adaptation to antiangiogenic treatment in glioblastoma. *Cancer Research, 72*(7), 1773–1783.

Huang, W., Metlakunta, A., Dedousis, N., Zhang, P., Sipula, I., Dube, J. J., et al. (2010). Depletion of liver Kupffer cells prevents the development of diet-induced hepatic steatosis and insulin resistance. *Diabetes, 59*(2), 347–357.

Hunter, K. W., Amin, R., Deasy, S., Ha, N. H., & Wakefield, L. (2018). Genetic insights into the morass of metastatic heterogeneity. *Nature Reviews. Cancer, 18*(4), 211–223.

Inoue, K., Nakada, K., Ogura, A., Isobe, K., Goto, Y., Nonaka, I., et al. (2000). Generation of mice with mitochondrial dysfunction by introducing mouse mtDNA carrying a deletion into zygotes. *Nature Genetics, 26*(2), 176–181.

Inoue-Yamauchi, A., & Oda, H. (2012). Depletion of mitochondrial fission factor DRP1 causes increased apoptosis in human colon cancer cells. *Biochemical and Biophysical Research Communications, 421*(1), 81–85.

Irwin, M. H., Johnson, L. W., & Pinkert, C. A. (1999). Isolation and microinjection of somatic cell-derived mitochondria and germline heteroplasmy in transmitochondrial mice. *Transgenic Research, 8*(2), 119–123.

Ishii, N., Fujii, M., Hartman, P. S., Tsuda, M., Yasuda, K., Senoo-Matsuda, N., et al. (1998). A mutation in succinate dehydrogenase cytochrome b causes oxidative stress and ageing in nematodes. *Nature, 394*(6694), 694–697.

Ishikawa, K., Takenaga, K., Akimoto, M., Koshikawa, N., Yamaguchi, A., Imanishi, H., et al. (2008). ROS-generating mitochondrial DNA mutations can regulate tumor cell metastasis. *Science, 320*(5876), 661–664.

Jo, A., Ham, S., Lee, G. H., Lee, Y. I., Kim, S., Lee, Y. S., et al. (2015). Efficient mitochondrial genome editing by CRISPR/Cas9. *BioMed Research International, 2015*, 305716.

Jones, R. A., Robinson, T. J., Liu, J. C., Shrestha, M., Voisin, V., Ju, Y., et al. (2016). RB1 deficiency in triple-negative breast cancer induces mitochondrial protein translation. *The Journal of Clinical Investigation, 126*(10), 3739–3757.

Jun, A. S., Trounce, I. A., Brown, M. D., Shoffner, J. M., & Wallace, D. C. (1996). Use of transmitochondrial cybrids to assign a complex I defect to the mitochondrial DNA-encoded NADH dehydrogenase subunit 6 gene mutation at nucleotide pair 14459 that causes Leber hereditary optic neuropathy and dystonia. *Molecular and Cellular Biology, 16*(3), 771–777.

Kashatus, J. A., Nascimento, A., Myers, L. J., Sher, A., Byrne, F. L., Hoehn, K. L., et al. (2015). Erk2 phosphorylation of Drp1 promotes mitochondrial fission and MAPK-driven tumor growth. *Molecular Cell, 57*(3), 537–551.

Kazuno, A. A., Munakata, K., Nagai, T., Shimozono, S., Tanaka, M., Yoneda, M., et al. (2006). Identification of mitochondrial DNA polymorphisms that alter mitochondrial matrix pH and intracellular calcium dynamics. *PLoS Genetics, 2*(8), e128.

Kenyon, L., & Moraes, C. T. (1997). Expanding the functional human mitochondrial DNA database by the establishment of primate xenomitochondrial cybrids. *Proceedings of the National Academy of Sciences of the United States of America, 94*(17), 9131–9135.

Kesterson, R. A., Johnson, L. W., Lambert, L. J., Vivian, J. L., Welch, D. R., & Ballinger, S. W. (2016). Generation of mitochondrial-nuclear eXchange mice via pronuclear transfer. *BioProtocols, 6*(20), e1976.

King, M. P., & Attardi, G. (1989). Human cells lacking mtDNA: Repopulation with exogenous mitochondria by complementation. *Science, 246*(4929), 500–503.

King, M. P., & Attardi, G. (1996). Isolation of human cell lines lacking mitochondrial DNA. *Methods in Enzymology, 264*, 304–313.

Krieger-Brauer, H. I., & Kather, H. (1992). Human fat cells possess a plasma membrane-bound H_2O_2-generating system that is activated by insulin via a mechanism bypassing the receptor kinase. *Journal of Clinical Investigation, 89*(3), 1006–1013.

Lagadinou, E. D., Sach, A., Callahan, K., Rossi, R. M., Neering, S. J., Minhajuddin, M., et al. (2013). BCL-2 inhibition targets oxidative phosphorylation and selectively eradicates quiescent human leukemia stem cells. *Cell Stem Cell, 12*(3), 329–341.

Lanthier, N., Molendi-Coste, O., Cani, P. D., van Rooijen, N., Horsmans, Y., & Leclercq, I. A. (2011). Kupffer cell depletion prevents but has no therapeutic effect on metabolic and inflammatory changes induced by a high-fat diet. *FASEB Journal, 25*(12), 4301–4311.

Latorre-Pellicer, A., Moreno-Loshuertos, R., Lechuga-Vieco, A. V., Sanchez-Cabo, F., Torroja, C., Acin-Perez, R., et al. (2016). Mitochondrial and nuclear DNA matching shapes metabolism and healthy ageing. *Nature, 535*(7613), 561–565.

Le Gal, K., Ibrahim, M. X., Wiel, C., Sayin, V. I., Akula, M. K., Karlsson, C., et al. (2015). Antioxidants can increase melanoma metastasis in mice. *Science Translational Medicine, 7*(308), 308re308.

Lee, S. R., Kwon, K. S., Kim, S. R., & Rhee, S. G. (1998). Reversible inactivation of protein-tyrosine phosphatase 1B in A431 cells stimulated with epidermal growth factor. *Journal of Biological Chemistry, 273*(25), 15366–15372.

Lee, K., Lee, M. H., Kang, Y. W., Rhee, K. J., Kim, T. U., & Kim, Y. S. (2012). Parkin induces apoptotic cell death in TNF-alpha-treated cervical cancer cells. *BMB Reports, 45*(9), 526–531.

Lee, H. C., Li, S. H., Lin, J. C., Wu, C. C., Yeh, D. C., & Wei, Y. H. (2004). Somatic mutations in the D-loop and decrease in the copy number of mitochondrial DNA in human hepatocellular carcinoma. *Mutation Research, 547*(1–2), 71–78.

Lee, J. H., Miele, M. E., Hicks, D. J., Phillips, K. K., Trent, J. M., Weissman, B. E., et al. (1996). KiSS-1, a novel human malignant melanoma metastasis-suppressor gene. *Journal of the National Cancer Institute, 88*(23), 1731–1737.

Lee, J. H., & Welch, D. R. (1997). Identification of highly expressed genes in metastasis-suppressed chromosome 6/human malignant melanoma hybrid cells using subtractive hybridization and differential display. *International Journal of Cancer, 71*(6), 1035–1044.

Lee, S. R., Yang, K. S., Kwon, J., Lee, C., Jeong, W., & Rhee, S. G. (2002). Reversible inactivation of the tumor suppressor PTEN by H_2O_2. *Journal of Biological Chemistry, 277*(23), 20336–20342.

Leslie, N. R., Bennett, D., Lindsay, Y. E., Stewart, H., Gray, A., & Downes, C. P. (2003). Redox regulation of PI 3-kinase signalling via inactivation of PTEN. *EMBO Journal, 22*(20), 5501–5510.

Levy, S. E., Waymire, K. G., Kim, Y. L., MacGregor, G. R., & Wallace, D. C. (1999). Transfer of chloramphenicol-resistant mitochondrial DNA into the chimeric mouse. *Transgenic Research, 8*(2), 137–145.

Li, F., Wang, Y., Zeller, K. I., Potter, J. J., Wonsey, D. R., O'Donnell, K. A., et al. (2005). Myc stimulates nuclearly encoded mitochondrial genes and mitochondrial biogenesis. *Molecular and Cellular Biology, 25*(14), 6225–6234.

Liemburg-Apers, D. C., Willems, P. H., Koopman, W. J., & Grefte, S. (2015). Interactions between mitochondrial reactive oxygen species and cellular glucose metabolism. *Archives of Toxicology, 89*(8), 1209–1226.

Lifsted, T., Le Voyer, T., Williams, M., Muller, W., Klein-Szanto, A., Buetow, K. H., et al. (1998). Identification of inbred mouse strains harboring genetic modifiers of mammary tumor age of onset and metastatic progression. *International Journal of Cancer*, 77(4), 640–644.

Liu, V. W., Wang, Y., Yang, H. J., Tsang, P. C., Ng, T. Y., Wong, L. C., et al. (2003). Mitochondrial DNA variant 16189T > C is associated with susceptibility to endometrial cancer. *Human Mutation*, 22(2), 173–174.

Lo, Y. Y., & Cruz, T. F. (1995). Involvement of reactive oxygen species in cytokine and growth factor induction of c-fos expression in chondrocytes. *Journal of Biological Chemistry*, 270(20), 11727–11730.

Lopez-Rios, F., Sanchez-Arago, M., Garcia-Garcia, E., Ortega, A. D., Berrendero, J. R., Pozo-Rodriguez, F., et al. (2007). Loss of the mitochondrial bioenergetic capacity underlies the glucose avidity of carcinomas. *Cancer Research*, 67(19), 9013–9017.

Lozoya, O. A., Martinez-Reyes, I., Wang, T., Grenet, D., Bushel, P., Li, J., et al. (2018). Mitochondrial nicotinamide adenine dinucleotide reduced (NADH) oxidation links the tricarboxylic acid (TCA) cycle with methionine metabolism and nuclear DNA methylation. *PLoS Biology*, 16(4), e2005707.

Lu, J., Sharma, L. K., & Bai, Y. (2009). Implications of mitochondrial DNA mutations and mitochondrial dysfunction in tumorigenesis. *Cell Research*, 19(7), 802–815.

Lu, J., Zheng, X., Li, F., Yu, Y., Chen, Z., Liu, Z., et al. (2017). Tunneling nanotubes promote intercellular mitochondria transfer followed by increased invasiveness in bladder cancer cells. *Oncotarget*, 8(9), 15539–15552.

Ma, J., Coarfa, C., Qin, X., Bonnen, P. E., Milosavljevic, A., Versalovic, J., et al. (2014). mtDNA haplogroup and single nucleotide polymorphisms structure human microbiome communities. *BMC Genomics*, 15, 257.

Ma, H., Folmes, C. D., Wu, J., Morey, R., Mora-Castilla, S., Ocampo, A., et al. (2015). Metabolic rescue in pluripotent cells from patients with mtDNA disease. *Nature*, 524(7564), 234–238.

Macintyre, A. N., Gerriets, V. A., Nichols, A. G., Michalek, R. D., Rudolph, M. C., Deoliveira, D., et al. (2014). The glucose transporter Glut1 is selectively essential for CD4 T cell activation and effector function. *Cell Metabolism*, 20(1), 61–72.

Macreadie, I. G., Novitski, C. E., Maxwell, R. J., John, U., Ooi, B. G., McMullen, G. L., et al. (1983). Biogenesis of mitochondria: The mitochondrial gene (aap1) coding for mitochondrial ATPase subunit 8 in Saccharomyces cerevisiae. *Nucleic Acids Research*, 11(13), 4435–4451.

Magee, R. G., Telonis, A. G., Loher, P., Londin, E., & Rigoutsos, I. (2018). Profiles of miRNA isoforms and tRNA fragments in prostate cancer. *Scientific Reports*, 8(1), 5314.

Marchington, D. R., Barlow, D., & Poulton, J. (1999). Transmitochondrial mice carrying resistance to chloramphenicol on mitochondrial DNA: Developing the first mouse model of mitochondrial DNA disease. *Nature Medicine*, 5(8), 957–960.

Markel, P., Shu, P., Ebeling, C., Carlson, G. A., Nagle, D. L., Smutko, J. S., et al. (1997). Theoretical and empirical issues for marker-assisted breeding of congenic mouse strains. *Nature Genetics*, 17(3), 280–284.

Martijn, J., Vosseberg, J., Guy, L., Offre, P., & Ettema, T. J. G. (2018). Deep mitochondrial origin outside the sampled alphaproteobacteria. *Nature*, 557(7703), 101–105.

Martinou, J. C., & Youle, R. J. (2011). Mitochondria in apoptosis: Bcl-2 family members and mitochondrial dynamics. *Developmental Cell*, 21(1), 92–101.

Matsuda, N., Sato, S., Shiba, K., Okatsu, K., Saisho, K., Gautier, C. A., et al. (2010). PINK1 stabilized by mitochondrial depolarization recruits Parkin to damaged mitochondria and activates latent Parkin for mitophagy. *Journal of Cell Biology*, 189(2), 211–221.

Mattiazzi, M., Vijayvergiya, C., Gajewski, C. D., DeVivo, D. C., Lenaz, G., Wiedmann, M., et al. (2004). The mtDNA T8993G (NARP) mutation results in an impairment of oxidative phosphorylation that can be improved by antioxidants. *Human Molecular Genetics*, *13*(8), 869–879.

Maynard, C. L., Elson, C. O., Hatton, R. D., & Weaver, C. T. (2012). Reciprocal interactions of the intestinal microbiota and immune system. *Nature*, *489*(7415), 231–241.

McKenzie, M., & Trounce, I. (2000). Expression of Rattus norvegicus mtDNA in Mus musculus cells results in multiple respiratory chain defects. *The Journal of Biological Chemistry*, *275*(40), 31514–31519.

McKenzie, M., Trounce, I. A., Cassar, C. A., & Pinkert, C. A. (2004). Production of homoplasmic xenomitochondrial mice. *Proceedings of the National Academy of Sciences of the United States of America*, *101*(6), 1685–1690.

Mittal, D., Gubin, M. M., Schreiber, R. D., & Smyth, M. J. (2014). New insights into cancer immunoediting and its three component phases—elimination, equilibrium and escape. *Current Opinion in Immunology*, *27*, 16–25.

Nagarajan, A., Malvi, P., & Wajapeyee, N. (2016). Oncogene-directed alterations in cancer cell metabolism. *Trends in Cancer*, *2*(7), 365–377.

Narendra, D. P., Jin, S. M., Tanaka, A., Suen, D. F., Gautier, C. A., Shen, J., et al. (2010). PINK1 is selectively stabilized on impaired mitochondria to activate Parkin. *PLoS Biology*, *8*(1), e1000298.

Noushmehr, H., Weisenberger, D. J., Diefes, K., Phillips, H. S., Pujara, K., Berman, B. P., et al. (2010). Identification of a CpG island methylator phenotype that defines a distinct subgroup of glioma. *Cancer Cell*, *17*(5), 510–522.

Okatsu, K., Koyano, F., Kimura, M., Kosako, H., Saeki, Y., Tanaka, K., et al. (2015). Phosphorylated ubiquitin chain is the genuine Parkin receptor. *Journal of Cell Biology*, *209*(1), 111–128.

Owens, K. M., Kulawiec, M., Desouki, M. M., Vanniarajan, A., & Singh, K. K. (2011). Impaired OXPHOS complex III in breast cancer. *PLoS One*, *6*(8), e23846.

Pagliarini, D. J., Calvo, S. E., Chang, B., Sheth, S. A., Vafai, S. B., Ong, S. E., et al. (2008). A mitochondrial protein compendium elucidates complex I disease biology. *Cell*, *134*(1), 112–123.

Park, H. J., Lyons, J. C., Ohtsubo, T., & Song, C. W. (1999). Acidic environment causes apoptosis by increasing caspase activity. *British Journal of Cancer*, *80*(12), 1892–1897.

Parrella, P., Xiao, Y., Fliss, M., Sanchez-Cespedes, M., Mazzarelli, P., Rinaldi, M., et al. (2001). Detection of mitochondrial DNA mutations in primary breast cancer and fine-needle aspirates. *Cancer Research*, *61*(20), 7623–7626.

Parsons, D. W., Jones, S., Zhang, X., Lin, J. C., Leary, R. J., Angenendt, P., et al. (2008). An integrated genomic analysis of human glioblastoma multiforme. *Science*, *321*(5897), 1807–1812.

Parsons, T. J., Muniec, D. S., Sullivan, K., Woodyatt, N., Alliston-Greiner, R., Wilson, M. R., et al. (1997). A high observed substitution rate in the human mitochondrial DNA control region. *Nature Genetics*, *15*(4), 363–368.

Pasquier, J., Guerrouahen, B. S., Al Thawadi, H., Ghiabi, P., Maleki, M., Abu-Kaoud, N., et al. (2013). Preferential transfer of mitochondria from endothelial to cancer cells through tunneling nanotubes modulates chemoresistance. *Journal of Translational Medicine*, *11*, 94.

Patananan, A. N., Wu, T. H., Chiou, P. Y., & Teitell, M. A. (2016). Modifying the mitochondrial genome. *Cell Metabolism*, *23*(5), 785–796.

Pavlides, S., Whitaker-Menezes, D., Castello-Cros, R., Flomenberg, N., Witkiewicz, A. K., Frank, P. G., et al. (2009). The reverse Warburg effect: Aerobic glycolysis in cancer associated fibroblasts and the tumor stroma. *Cell Cycle*, *8*(23), 3984–4001.

Pello, R., Martin, M. A., Carelli, V., Nijtmans, L. G., Achilli, A., Pala, M., et al. (2008). Mitochondrial DNA background modulates the assembly kinetics of OXPHOS complexes in a cellular model of mitochondrial disease. *Human Molecular Genetics*, *17*(24), 4001–4011.

Pfefferle, A. D., Herschkowitz, J. I., Usary, J., Harrell, J. C., Spike, B. T., Adams, J. R., et al. (2013). Transcriptomic classification of genetically engineered mouse models of breast cancer identifies human subtype counterparts. *Genome Biology*, *14*(11), R125.
Pfeiffer, T., Schuster, S., & Bonhoeffer, S. (2001). Cooperation and competition in the evolution of ATP-producing pathways. *Science*, *292*(5516), 504–507.
Phillips, K. K., Welch, D. R., Miele, M. E., Lee, J. H., Wei, L. L., & Weissman, B. E. (1996). Suppression of MDA-MB-435 breast carcinoma cell metastasis following the introduction of human chromosome 11. *Cancer Research*, *56*(6), 1222–1227.
Picard, M., Zhang, J., Hancock, S., Derbeneva, O., Golhar, R., Golik, P., et al. (2014). Progressive increase in mtDNA 3243A>G heteroplasmy causes abrupt transcriptional reprogramming. *Proceedings of the National Academy of Sciences of the United States of America*, *111*(38), E4033–E4042.
Pinkert, C. A., Irwin, M. H., Johnson, L. W., & Moffatt, R. J. (1997). Mitochondria transfer into mouse ova by microinjection. *Transgenic Research*, *6*(6), 379–383.
Piskounova, E., Agathocleous, M., Murphy, M. M., Hu, Z., Huddlestun, S. E., Zhao, Z., et al. (2015). Oxidative stress inhibits distant metastasis by human melanoma cells. *Nature*, *527*(7577), 186–191.
Porter, D. L., Levine, B. L., Kalos, M., Bagg, A., & June, C. H. (2011). Chimeric antigen receptor-modified T cells in chronic lymphoid leukemia. *New England Journal of Medicine*, *365*(8), 725–733.
Reddy, P., Ocampo, A., Suzuki, K., Luo, J., Bacman, S. R., Williams, S. L., et al. (2015). Selective elimination of mitochondrial mutations in the germline by genome editing. *Cell*, *161*(3), 459–469.
Redza-Dutordoir, M., & Averill-Bates, D. A. (2016). Activation of apoptosis signalling pathways by reactive oxygen species. *Biochimica et Biophysica Acta*, *1863*(12), 2977–2992.
Rehman, J., Zhang, H. J., Toth, P. T., Zhang, Y., Marsboom, G., Hong, Z., et al. (2012). Inhibition of mitochondrial fission prevents cell cycle progression in lung cancer. *FASEB Journal*, *26*(5), 2175–2186.
Renault, T. T., Floros, K. V., Elkholi, R., Corrigan, K. A., Kushnareva, Y., Wieder, S. Y., et al. (2015). Mitochondrial shape governs BAX-induced membrane permeabilization and apoptosis. *Molecular Cell*, *57*(1), 69–82.
Rivera, C. A., Adegboyega, P., van Rooijen, N., Tagalicud, A., Allman, M., & Wallace, M. (2007). Toll-like receptor-4 signaling and Kupffer cells play pivotal roles in the pathogenesis of non-alcoholic steatohepatitis. *Journal of Hepatology*, *47*(4), 571–579.
Ro, S., Ma, H. Y., Park, C., Ortogero, N., Song, R., Hennig, G. W., et al. (2013). The mitochondrial genome encodes abundant small noncoding RNAs. *Cell Research*, *23*(6), 759–774.
Robin, E. D., & Wong, R. (1988). Mitochondrial DNA molecules and virtual number of mitochondria per cell in mammalian cells. *Journal of Cellular Physiology*, *136*(3), 507–513.
Ross, O. A., McCormack, R., Curran, M. D., Duguid, R. A., Barnett, Y. A., Rea, I. M., et al. (2001). Mitochondrial DNA polymorphism: Its role in longevity of the Irish population. *Experimental Gerontology*, *36*(7), 1161–1178.
Rossignol, R., Gilkerson, R., Aggeler, R., Yamagata, K., Remington, S. J., & Capaldi, R. A. (2004). Energy substrate modulates mitochondrial structure and oxidative capacity in cancer cells. *Cancer Research*, *64*(3), 985–993.
Routy, B., Le Chatelier, E., Derosa, L., Duong, C. P. M., Alou, M. T., Daillere, R., et al. (2018). Gut microbiome influences efficacy of PD-1-based immunotherapy against epithelial tumors. *Science*, *359*(6371), 91–97.
Sallmyr, A., Fan, J., Datta, K., Kim, K. T., Grosu, D., Shapiro, P., et al. (2008). Internal tandem duplication of FLT3 (FLT3/ITD) induces increased ROS production, DNA damage, and misrepair: Implications for poor prognosis in AML. *Blood*, *111*(6), 3173–3182.
Santel, A., & Fuller, M. T. (2001). Control of mitochondrial morphology by a human mitofusin. *Journal of Cell Science*, *114*(Pt. 5), 867–874.

Sarraf, S. A., Raman, M., Guarani-Pereira, V., Sowa, M. E., Huttlin, E. L., Gygi, S. P., et al. (2013). Landscape of the PARKIN-dependent ubiquitylome in response to mitochondrial depolarization. *Nature*, *496*(7445), 372–376.

Sassera, D., Lo, N., Epis, S., D'Auria, G., Montagna, M., Comandatore, F., et al. (2011). Phylogenomic evidence for the presence of a flagellum and cbb(3) oxidase in the free-living mitochondrial ancestor. *Molecular Biology and Evolution*, *28*(12), 3285–3296.

Schreck, R., Rieber, P., & Baeuerle, P. A. (1991). Reactive oxygen intermediates as apparently widely used messengers in the activation of the NF-kappa B transcription factor and HIV-1. *EMBO Journal*, *10*(8), 2247–2258.

Semenza, G. L. (2013). Cancer-stromal cell interactions mediated by hypoxia-inducible factors promote angiogenesis, lymphangiogenesis, and metastasis. *Oncogene*, *32*(35), 4057–4063.

Senoo-Matsuda, N., Yasuda, K., Tsuda, M., Ohkubo, T., Yoshimura, S., Nakazawa, H., et al. (2001). A defect in the cytochrome b large subunit in complex II causes both superoxide anion overproduction and abnormal energy metabolism in *Caenorhabditis elegans*. *The Journal of Biological Chemistry*, *276*(45), 41553–41558.

Seo, J. H., Agarwal, E., Bryant, K. G., Caino, M. C., Kim, E. T., Kossenkov, A. V., et al. (2018). Syntaphilin ubiquitination regulates mitochondrial dynamics and tumor cell movements. *Cancer Research*, *78*(15), 4215–4228.

Seraj, M. J., Samant, R. S., Verderame, M. F., & Welch, D. R. (2000). Functional evidence for a novel human breast carcinoma metastasis suppressor, BRMS1, encoded at chromosome 11q13. *Cancer Research*, *60*(11), 2764–2769.

Serasinghe, M. N., Wieder, S. Y., Renault, T. T., Elkholi, R., Asciolla, J. J., Yao, J. L., et al. (2015). Mitochondrial division is requisite to RAS-induced transformation and targeted by oncogenic MAPK pathway inhibitors. *Molecular Cell*, *57*(3), 521–536.

Shuster, R. C., Rubenstein, A. J., & Wallace, D. C. (1988). Mitochondrial DNA in anucleate human blood cells. *Biochemical and Biophysical Research Communications*, *155*(3), 1360–1365.

Shutt, T., Geoffrion, M., Milne, R., & McBride, H. M. (2012). The intracellular redox state is a core determinant of mitochondrial fusion. *EMBO Reports*, *13*(10), 909–915.

Singh, K. K., Ayyasamy, V., Owens, K. M., Koul, M. S., & Vujcic, M. (2009). Mutations in mitochondrial DNA polymerase-gamma promote breast tumorigenesis. *Journal of Human Genetics*, *54*(9), 516–524.

Sligh, J. E., Levy, S. E., Waymire, K. G., Allard, P., Dillehay, D. L., Nusinowitz, S., et al. (2000). Maternal germ-line transmission of mutant mtDNAs from embryonic stem cell-derived chimeric mice. *Proceedings of the National Academy of Sciences of the United States of America*, *97*(26), 14461–14466.

Smigrodzki, R. M., & Khan, S. M. (2005). Mitochondrial microheteroplasmy and a theory of aging and age-related disease. *Rejuvenation Research*, *8*(3), 172–198.

Smiraglia, D. J., Kulawiec, M., Bistulfi, G. L., Gupta, S. G., & Singh, K. K. (2008). A novel role for mitochondria in regulating epigenetic modification in the nucleus. *Cancer Biology & Therapy*, *7*(8), 1182–1190.

Smirnova, E., Shurland, D. L., Ryazantsev, S. N., & van der Bliek, A. M. (1998). A human dynamin-related protein controls the distribution of mitochondria. *Journal of Cell Biology*, *143*(2), 351–358.

Sripada, L., Tomar, D., Prajapati, P., Singh, R., Singh, A. K., & Singh, R. (2012). Systematic analysis of small RNAs associated with human mitochondria by deep sequencing: Detailed analysis of mitochondrial associated miRNA. *PLoS One*, *7*(9), e44873.

Steeg, P. S., & Theodorescu, D. (2007). Metastasis: A therapeutic target for cancer. *Nature Clinical Practice Oncology*, *5*(4), 206–219.

Stewart, J. B., Freyer, C., Elson, J. L., Wredenberg, A., Cansu, Z., Trifunovic, A., et al. (2008). Strong purifying selection in transmission of mammalian mitochondrial DNA. *PLoS Biology*, *6*(1), e10.

Strasser, A., Harris, A. W., Bath, M. L., & Cory, S. (1990). Novel primitive lymphoid tumours induced in transgenic mice by cooperation between myc and bcl-2. *Nature*, *348*(6299), 331–333.
Suganuma, K., Miwa, H., Imai, N., Shikami, M., Gotou, M., Goto, M., et al. (2010). Energy metabolism of leukemia cells: Glycolysis versus oxidative phosphorylation. *Leukemia and Lymphoma*, *51*(11), 2112–2119.
Sundaresan, M., Yu, Z. X., Ferrans, V. J., Irani, K., & Finkel, T. (1995). Requirement for generation of H_2O_2 for platelet-derived growth factor signal transduction. *Science*, *270*(5234), 296–299.
Swerdlow, R. H., Parks, J. K., Davis, J. N., 2nd, Cassarino, D. S., Trimmer, P. A., Currie, L. J., et al. (1998). Matrilineal inheritance of complex I dysfunction in a multigenerational Parkinson's disease family. *Annals of Neurology*, *44*(6), 873–881.
Swerdlow, R. H., Parks, J. K., Miller, S. W., Tuttle, J. B., Trimmer, P. A., Sheehan, J. P., et al. (1996). Origin and functional consequences of the complex I defect in Parkinson's disease. *Annals of Neurology*, *40*(4), 663–671.
Taanman, J. W. (1999). The mitochondrial genome: Structure, transcription, translation and replication. *Biochimica et Biophysica Acta*, *1410*(2), 103–123.
Tan, A. S., Baty, J. W., Dong, L. F., Bezawork-Geleta, A., Endaya, B., Goodwin, J., et al. (2015). Mitochondrial genome acquisition restores respiratory function and tumorigenic potential of cancer cells without mitochondrial DNA. *Cell Metabolism*, *21*(1), 81–94.
Tao, S., Luo, Y., Bin, H., Liu, J., Qian, X., Ni, Y., et al. (2016). Paraoxonase 2 modulates a proapoptotic function in LS174T cells in response to quorum sensing molecule N-(3-oxododecanoyl)-L-homoserine lactone. *Scientific Reports*, *6*, 28778.
Tay, S. P., Yeo, C. W., Chai, C., Chua, P. J., Tan, H. M., Ang, A. X., et al. (2010). Parkin enhances the expression of cyclin-dependent kinase 6 and negatively regulates the proliferation of breast cancer cells. *Journal of Biological Chemistry*, *285*(38), 29231–29238.
Telonis, A. G., Loher, P., Honda, S., Jing, Y., Palazzo, J., Kirino, Y., et al. (2015). Dissecting tRNA-derived fragment complexities using personalized transcriptomes reveals novel fragment classes and unexpected dependencies. *Oncotarget*, *6*(28), 24797–24822.
Telonis, A. G., & Rigoutsos, I. (2018). Race disparities in the contribution of miRNA isoforms and tRNA-derived fragments to triple-negative breast cancer. *Cancer Research*, *78*(5), 1140–1154.
Topalian, S. L., Hodi, F. S., Brahmer, J. R., Gettinger, S. N., Smith, D. C., McDermott, D. F., et al. (2012). Safety, activity, and immune correlates of anti-PD-1 antibody in cancer. *New England Journal of Medicine*, *366*(26), 2443–2454.
Tourmente, M., Hirose, M., Ibrahim, S., Dowling, D. K., Tompkins, D. M., Roldan, E. R. S., et al. (2017). mtDNA polymorphism and metabolic inhibition affect sperm performance in conplastic mice. *Reproduction*, *154*(4), 341–354.
Trifunovic, A., Wredenberg, A., Falkenberg, M., Spelbrink, J. N., Rovio, A. T., Bruder, C. E., et al. (2004). Premature ageing in mice expressing defective mitochondrial DNA polymerase. *Nature*, *429*(6990), 417–423.
Tsujimoto, Y., Finger, L. R., Yunis, J., Nowell, P. C., & Croce, C. M. (1984). Cloning of the chromosome breakpoint of neoplastic B cells with the t(14;18) chromosome translocation. *Science*, *226*(4678), 1097–1099.
van der Walt, J. M., Nicodemus, K. K., Martin, E. R., Scott, W. K., Nance, M. A., Watts, R. L., et al. (2003). Mitochondrial polymorphisms significantly reduce the risk of Parkinson disease. *American Journal of Human Genetics*, *72*(4), 804–811.
Vander Heiden, M. G., & DeBerardinis, R. J. (2017). Understanding the intersections between metabolism and cancer biology. *Cell*, *168*(4), 657–669.
Vetizou, M., Pitt, J. M., Daillere, R., Lepage, P., Waldschmitt, N., Flament, C., et al. (2015). Anticancer immunotherapy by CTLA-4 blockade relies on the gut microbiota. *Science*, *350*(6264), 1079–1084.

Viale, A., Pettazzoni, P., Lyssiotis, C. A., Ying, H., Sanchez, N., Marchesini, M., et al. (2014). Oncogene ablation-resistant pancreatic cancer cells depend on mitochondrial function. *Nature, 514*(7524), 628–632.

Viaud, S., Saccheri, F., Mignot, G., Yamazaki, T., Daillere, R., Hannani, D., et al. (2013). The intestinal microbiota modulates the anticancer immune effects of cyclophosphamide. *Science, 342*(6161), 971–976.

Vivian, C. J., Brinker, A. E., Graw, S., Koestler, D. C., Legendre, C., Gooden, G. C., et al. (2017). Mitochondrial genomic backgrounds affect nuclear DNA methylation and gene expression. *Cancer Research, 77*(22), 6202–6214.

von Eyss, B., Jaenicke, L. A., Kortlever, R. M., Royla, N., Wiese, K. E., Letschert, S., et al. (2015). A MYC-driven change in mitochondrial dynamics limits YAP/TAZ function in mammary epithelial cells and breast cancer. *Cancer Cell, 28*(6), 743–757.

Vyas, S., Zaganjor, E., & Haigis, M. C. (2016). Mitochondria and cancer. *Cell, 166*(3), 555–566.

Wallace, D. C. (2012). Mitochondria and cancer. *Nature Reviews. Cancer, 12*(10), 685–698.

Wallace, D. C. (2015). Mitochondrial DNA variation in human radiation and disease. *Cell, 163*(1), 33–38.

Wallace, D. C. (2016). Genetics: Mitochondrial DNA in evolution and disease. *Nature, 535*(7613), 498–500.

Wallace, D. C., Bunn, C. L., & Eisenstadt, J. M. (1975). Cytoplasmic transfer of chloramphenicol resistance in human tissue culture cells. *Journal of Cell Biology, 67*(1), 174–188.

Wallace, D. C., & Chalkia, D. (2013). Mitochondrial DNA genetics and the heteroplasmy conundrum in evolution and disease. *Cold Spring Harbor Perspectives in Biology, 5*(11), a021220.

Wan, Y. Y., Zhang, J. F., Yang, Z. J., Jiang, L. P., Wei, Y. F., Lai, Q. N., et al. (2014). Involvement of Drp1 in hypoxia-induced migration of human glioblastoma U251 cells. *Oncology Reports, 32*(2), 619–626.

Wang, T., Cai, G., Qiu, Y., Fei, N., Zhang, M., Pang, X., et al. (2012). Structural segregation of gut microbiota between colorectal cancer patients and healthy volunteers. *The ISME Journal, 6*(2), 320–329.

Wang, Z., & Wu, M. (2015). An integrated phylogenomic approach toward pinpointing the origin of mitochondria. *Scientific Reports, 5*, 7949.

Warburg, O. (1956a). On respiratory impairment in cancer cells. *Science, 124*(3215), 269–270.

Warburg, O. (1956b). On the origin of cancer cells. *Science, 123*(3191), 309–314.

Warburg, O., Wind, F., & Negelein, E. (1927). The metabolism of tumors in the body. *Journal of General Physiology, 8*(6), 519–530.

Weinstein, I. B., & Joe, A. K. (2006). Mechanisms of disease: Oncogene addiction–a rationale for molecular targeting in cancer therapy. *Nature Clinical Practice. Oncology, 3*(8), 448–457.

Welch, D. R., Chen, P., Miele, M. E., McGary, C. T., Bower, J. M., Stanbridge, E. J., et al. (1994). Microcell-mediated transfer of chromosome 6 into metastatic human C8161 melanoma cells suppresses metastasis but does not inhibit tumorigenicity. *Oncogene, 9*(1), 255–262.

Wellen, K. E., Hatzivassiliou, G., Sachdeva, U. M., Bui, T. V., Cross, J. R., & Thompson, C. B. (2009). ATP-citrate lyase links cellular metabolism to histone acetylation. *Science, 324*(5930), 1076–1080.

Whitaker-Menezes, D., Martinez-Outschoorn, U. E., Flomenberg, N., Birbe, R. C., Witkiewicz, A. K., Howell, A., et al. (2011). Hyperactivation of oxidative mitochondrial metabolism in epithelial cancer cells in situ: Visualizing the therapeutic effects of metformin in tumor tissue. *Cell Cycle, 10*(23), 4047–4064.

Wikstrom, J. D., Mahdaviani, K., Liesa, M., Sereda, S. B., Si, Y., Las, G., et al. (2014). Hormone-induced mitochondrial fission is utilized by brown adipocytes as an amplification pathway for energy expenditure. *EMBO Journal*, *33*(5), 418–436.

Wilkins, H. M., Carl, S. M., & Swerdlow, R. H. (2014). Cytoplasmic hybrid (cybrid) cell lines as a practical model for mitochondriopathies. *Redox Biology*, *2*, 619–631.

Wiseman, H., & Halliwell, B. (1996). Damage to DNA by reactive oxygen and nitrogen species: Role in inflammatory disease and progression to cancer. *Biochemical Journal*, *313*(Pt. 1), 17–29.

Xie, C. H., Naito, A., Mizumachi, T., Evans, T. T., Douglas, M. G., Cooney, C. A., et al. (2007). Mitochondrial regulation of cancer associated nuclear DNA methylation. *Biochemical and Biophysical Research Communications*, *364*(3), 656–661.

Yamaoka, M., Isobe, K., Shitara, H., Yonekawa, H., Miyabayashi, S., & Hayashi, J. I. (2000). Complete repopulation of mouse mitochondrial DNA-less cells with rat mitochondrial DNA restores mitochondrial translation but not mitochondrial respiratory function. *Genetics*, *155*(1), 301–307.

Yang, D., Oyaizu, Y., Oyaizu, H., Olsen, G. J., & Woese, C. R. (1985). Mitochondrial origins. *Proceedings of the National Academy of Sciences of the United States of America*, *82*(13), 4443–4447.

Yu, X., Gimsa, U., Wester-Rosenlof, L., Kanitz, E., Otten, W., Kunz, M., et al. (2009). Dissecting the effects of mtDNA variations on complex traits using mouse conplastic strains. *Genome Research*, *19*(1), 159–165.

Yu, X., Koczan, D., Sulonen, A. M., Akkad, D. A., Kroner, A., Comabella, M., et al. (2008). mtDNA nt13708A variant increases the risk of multiple sclerosis. *PLoS One*, *3*(2), e1530.

Zhang, W., Bojorquez-Gomez, A., Velez, D. O., Xu, G., Sanchez, K. S., Shen, J. P., et al. (2018). A global transcriptional network connecting noncoding mutations to changes in tumor gene expression. *Nature Genetics*, *50*(4), 613–620.

Zhang, Q., Raoof, M., Chen, Y., Sumi, Y., Sursal, T., Junger, W., et al. (2010). Circulating mitochondrial DAMPs cause inflammatory responses to injury. *Nature*, *464*(7285), 104–107.

Zhao, J., Zhang, J., Yu, M., Xie, Y., Huang, Y., Wolff, D. W., et al. (2013). Mitochondrial dynamics regulates migration and invasion of breast cancer cells. *Oncogene*, *32*(40), 4814–4824.

CHAPTER FOUR

Pathways- and epigenetic-based assessment of relative immune infiltration in various types of solid tumors

Manny D. Bacolod[a,]*, Francis Barany[a], Karsten Pilones[b], Paul B. Fisher[c], Romulo J. de Castro[d]

[a]Department of Microbiology and Immunology, Weill Cornell Medicine, New York, NY, United States
[b]Department of Radiation Oncology, Weill Cornell Medicine, New York, NY, United States
[c]Department of Human and Molecular Genetics, VCU Institute of Molecular Medicine, VCU Massey Cancer Center, Virginia Commonwealth University, School of Medicine, Richmond, VA, United States
[d]3R Biosystems, Long Beach, CA, United States
*Corresponding author: e-mail address: mdb2005@med.cornell.edu

Contents

1. Introduction	108
2. Methodology	110
2.1 Genomic datasets and databases	110
2.2 Statistical and analytical tools	111
3. Results	114
3.1 Gene set enrichment analyses (GSEA) indicate differences in the extent of relative activation of T cells and other immune signaling pathways, in different types of cancer	114
3.2 Renal, skin, head and neck, and esophageal cancers exhibit the most highly enriched pathways related to activation and proliferation of helper and cytotoxic T cells	118
3.3 Relative to normal tissues, T cell-related pathways are not significantly enriched in the primary tumors of lung, colorectal, liver and bile duct cancers	120
3.4 For certain cancer types, immune infiltration/activation correlates with improved prognosis	121
3.5 Identification of CpG promoter sites most highly associated with expression of important T lymphocyte marker genes	128
3.6 Promoter CpG methylation likely influences the binding of modified histones, and thereby the transcription of immune-related genes	131
4. Discussion	133
5. Expert opinion	138
Acknowledgments	139
References	139

Abstract

Recent clinical studies document the power of immunotherapy in treating subsets of patients with advanced cancers. In this context and with multiple cancer immunotherapeutics already evaluated in the clinic and a large number in various stages of clinical trials, it is imperative to comprehensively examine genomics data to better comprehend the role of immunity in different cancers in predicting response to therapy and in directing appropriate therapies. The approach we chose is to scrutinize the pathways and epigenetic factors predicted to drive immune infiltration in different cancer types using publicly available TCGA transcriptional and methylation datasets, along with accompanying clinico-pathological data. We observed that the relative activation of T cells and other immune signaling pathways differs across cancer types. For example, pathways related to activation and proliferation of helper and cytotoxic T cells appear to be more highly enriched in kidney, skin, head and neck, and esophageal cancers compared to those of lung, colorectal, and liver or bile duct cancers. The activation of these immune-related pathways positively associated with prognosis in certain cancer types, most notably melanoma, head and neck, and cervical cancers. Integrated methylation and expression data (along with publicly available, ENCODE-generated histone ChIP Seq and DNAse hypersensitivity data) predict that epigenetic regulation is a primary factor driving transcriptional activation of a number of genes crucial to immunity in cancer, including T cell receptor genes (e.g., *CD3D*, *CD3E*), *CTLA4*, and *GZMA*. However, the extent to which epigenetic factors (primarily methylation at promoter regions) affect transcription of immune-related genes may vary across cancer types. For example, there is a high negative correlation between promoter CpG methylation and CD3D expression in renal and thyroid cancers, but not in brain tumors. The types of analyses we have undertaken provide insights into the relationships between immune modulation and cancer etiology and progression, offering clues into ways of therapeutically manipulating the immune system to promote immune recognition and immunotherapy.

1. Introduction

Modern genomics has profoundly influenced our current appreciation of many areas of cancer research, including cancer genetics, therapeutics, diagnostics, biology, and epidemiology (Berger & Mardis, 2018; Nguyen & Gocke, 2017; Ning et al., 2014; Senft, Leiserson, Ruppin, & Ronai, 2017; Stadler, Schrader, Vijai, Robson, & Offit, 2014). Genome-wide molecular profiling (e.g., copy number, methylation, mutation, expression levels of mRNAs and non-coding RNAs) of cancer specimens, using primarily microarray- and sequencing-based tools, has permitted a comprehensive stratification of tumors, identification of predictive biomarkers and potential therapeutic targets, among many other discoveries. More importantly, there are now tens of thousands of cancer-related genomic datasets that are available to the public such as those generated by The Cancer Genome Atlas (TCGA) (Kaiser, 2005), ENCODE (ENCODE Project Consortium, 2004), and those which can be found at data repositories such as Gene Expression Omnibus (GEO) (Barrett & Edgar, 2006). Cancer immunology, a field that is technically dominated by immunohistochemistry and flow cytometry (Bondanza & Casucci, 2016; Ursini-Siegel & Beauchemin, 2016), has also benefited positively from genomics (and public datasets), especially in recent years. Understanding immune infiltration in cancer through the use of genome-wide analytical tools may have been unintentionally hindered by the quest for highly homogeneous tumor samples prior to molecular profiling (Glanzer & Eberwine, 2004; Player, Barrett, & Kawasaki, 2004). However, we are currently witnessing a rapid expansion in the development of cancer immunotherapy. Between January 1, 2016 and April 16, 2018, 25 out of 93 FDA approvals for cancer drugs, related diagnostics, and safety notifications involved immunotherapeutic monoclonal antibodies (mAbs) against solid tumors (https://www.fda.gov/Drugs/InformationOnDrugs/ApprovedDrugs/).

There is a great interest in employing genomic tools in order to learn more about the interactions between cancer and immune cells. One particular study has examined molecular markers of immune infiltration in cancer by associating T cells' cytolytic activity (CYT), defined as geometric mean of expression levels of granzyme A (GZMA) and perforin 1 (PRF1), with genome-wide transcription in tumor samples (Rooney, Shukla, Wu, Getz, & Hacohen, 2015). Other reports described the presence of various immune cell infiltrates by examining the gene expression signatures for T cells, CD8 T cells, B cells, macrophages, and NK, Mono, MFm2, TregCD8, and NKCD8 cells (Danaher et al., 2017; Iglesia et al., 2016; Li et al., 2016; Senbabaoglu et al., 2016; Siemers et al., 2017). These observations based on mRNA profiles of tumors are generally consistent with our understanding of immune infiltration in cancer based on immunohistochemical studies, mainly that various immune infiltrating cells, as well as stromal, lymphatic, and myeloid cells, are crucial to the proliferation of tumors originating from epithelial cells (Giraldo et al., 2014). Among the infiltrating immune cells widely analyzed for immunohistochemical detection include those for CD4 T cells, CD8 T cells, B cells (CD20), macrophages (CD68) and dendritic cells (CD11C) (Degnim et al., 2017). Immunohistochemistry has also demonstrated that T cell surface markers such as CD3, CD4, and CD8 are associated with better prognosis in melanoma, head and neck, and bladder cancer, while at least one of these markers is positively associated with prognosis in colorectal, breast, ovarian, lung, urothelial, endometrial, esophageal, liver, and pancreatic cancers (Bethmann, Feng, & Fox, 2017; Fridman, Pages, Sautes-Fridman, & Galon, 2012). Another marker associated with good prognosis is CD45O (demonstrated in colorectal, ovarian, endometrial, and liver cancers). Immunohistochemical staining has also shown that FOXP3 (a marker of T reg cells) is negatively associated with prognosis in bladder, endometrial, liver, pancreatic, glioblastoma multiforme, and renal cancer, but reflects a favorable response in colorectal and breast cancers.

Despite the vast knowledge accrued in recent years, there are still many aspects of cancer immune infiltration (and the related field of cancer immunotherapy) that may be uncovered through analysis of genomics and accompanying clinico-pathological data of tumor samples. Another instructive area is the use of molecular profiling to identify predictive markers, and define resistance mechanisms for the FDA-approved immune checkpoint inhibitory drugs now being used in the clinic, such as those targeting CTLA4 (ipilimumab), PD1 (nivolumab, pembrolizumab), and PDL1 (atezolizumab) (Couzin-Frankel, 2013; D'Errico, Machado, & Sainz, 2017; Greil, Hutterer, Hartmann, & Pleyer, 2017; Sukari, Nagasaka, Al-Hadidi, & Lum, 2016; Sweis & Luke, 2017). For this manuscript, we have employed bioinformatic tools to analyze the differences and similarities among different types of solid tumors with respect to activation of immune-related pathways, and the predicted epigenetic regulation of genes that are crucial to immune cell infiltration.

2. Methodology

2.1 Genomic datasets and databases

2.1.1 TCGA expression and methylation datasets

For our analysis we have employed various mRNA expression (generated using Illumina HiSeq 2000) and CpG methylation (using Illumina Infinium 450K BeadChip) datasets generated by The Cancer Genome Atlas (TCGA) (https://tcga-data.nci.nih.gov/tcga/) project (Kaiser, 2005). The array-formatted datasets, along with clinical annotations, were obtained from the UCSC Cancer Genomics website (https://genome-cancer.ucsc.edu/) (Goldman et al., 2013; Zhu et al., 2009). The TCGA datasets represent *32 different solid cancer cohorts including*: adrenocortical carcinoma [ACC], bladder urothelial carcinoma [BLCA], breast invasive carcinoma [BRCA], cervical squamous cell carcinoma and endocervical adenocarcinoma [CESC], cholangiocarcinoma [CHOL], colon adenocarcinoma [COAD], lymphoid neoplasm diffuse large B-cell lymphoma [DLBC], esophageal carcinoma [ESCA], glioblastoma multiforme [GBM], head and neck squamous cell carcinoma [HNSC], kidney chromophobe [KICH], kidney renal clear cell carcinoma [KIRC], kidney renal papillary cell carcinoma [KIRP], brain lower grade glioma [LGG], liver hepatocellular carcinoma [LIHC], lung adenocarcinoma [LUAD], lung squamous cell carcinoma [LUSC], mesothelioma [MESO], ovarian serous cystadenocarcinoma [OV], pancreatic adenocarcinoma [PAAD], pheochromocytoma and paraganglioma [PCPG], prostate adenocarcinoma [PRAD], rectum adenocarcinoma [READ], sarcoma [SARC], skin cutaneous melanoma [SKCM], stomach adenocarcinoma [STAD], testicular germ cell tumors [TGCT], thymoma [THYM], thyroid carcinoma [THCA], uterine corpus endometrial carcinoma [UCEC], uterine carcinosarcoma [UCS], and uveal melanoma [UVM].

2.1.2 Additional methylation datasets

Also downloaded from Gene Expression Omnibus (GEO) are the Illumina 450K datasets: GSE40699, encompassing 63 cell lines which are part of the ENCODE Project, and generated by R. M. Myers' laboratory (Hudson Alpha Inst.); and GSE68379 (Iorio et al., 2016), generated by the Esteller group, and covers 1001 cancer cell lines. However, only select information from these datasets was relevant for our current analysis.

2.2 Statistical and analytical tools

A number of genomic and statistical tools were employed for our current investigation (as described in the following).

2.2.1 Basic statistical processing and analyses of public genomic datasets

All basic statistical analyses (comparative statistics, normalization, correlation analyses, hierarchical clustering, etc.) were performed using JMP Pro 11/JMP Genomics software (SAS, Cary, NC), and Gene-E (Broad Institute, Cambridge, MA).

2.2.2 Pathways prediction

Prediction of associated molecular pathways was conducted using Gene Set Enrichment Analysis (GSEA) available through the Broad Institute (www.broadinstitute.org/gsea/) (Subramanian et al., 2005). GSEA starts with the recognition that genes associate in particular groups (or gene sets), representing pathways and functionalities, such as those defined in Biocarta (http://software.broadinstitute.org/gsea/msigdb/genesets.jsp?collection=CP:BIOCARTA) (Note: registration is required in order to access the links to these pathways). The program was set to examine the collective trends of how members of each of the 184 Biocarta gene sets (i.e., those with at least 12 member genes) behave through the calculation of an Enrichment Score (ES). The input data for the analyses are genome-wide mRNA expression data for the primary tumor (PT) compared with solid normal (SN) dataset, except for the SKCM dataset which lacked SN samples (thus the input data are for the PT compared with metastasis samples instead) (see Fig. 1 for the flowchart). However, only 18 of the 32 TCGA cohorts have substantial numbers (at least 9) of baseline subsets (i.e., solid normal for most cohorts, and primary tumors for SKCM), and have been included in this particular analysis. These cohorts were BLCA, BRCA, CHOL, COAD, ESCA, HNSC, KICH, KIRC, KIRP, LIHC, LUAD, LUSC, PRAD, READ, SKCM, STAD, THCA, and UCEC. A high concentration of highly ranked genes (rank=1 for the gene with the highest PT vs. SN weighted signal-to-noise ratio) belonging to a particular gene set translates to an ES value close to 1, which may be interpreted as that gene set being

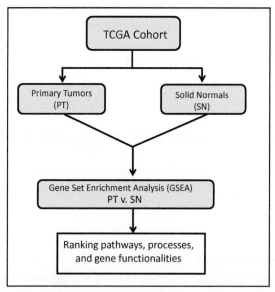

Fig. 1 The scheme used to rank the Biocarta pathways activated in primary tumors vs. solid normal (or metastasis relative to solid normal in the case of SKCM) across the 18 TCGA cohorts.

likely associated with upregulation in PT relative to SN. An ES value close to −1 may likewise be interpreted as that gene set being likely associated with downregulation.

2.2.3 Survival analyses

Survival analyses were performed using modules included in BRB-ArrayTools, an R-based Excel add-in (Simon et al., 2007). Using as input the RNASeq-generated total mRNA expression data of primary tumors and the corresponding follow-up records (follow-up time and vital status), Cox's proportional hazards model (Cox, 1972) was used to predict the prognostic value of each gene. In addition, the survival risk of a group of genes (again, such as those listed in Biocarta) was calculated using Efron and Tibshirani's gene set analysis (GSA) summarization method (Efron & Tibshirani, 2007) incorporated in BRB Array Tools.

2.2.4 Predicting the epigenetic regulation of genes important in T cell-related pathways

Whether epigenetics play important roles in the transcriptional activation of crucial T cell marker genes can be ascertained through a series of analyses: (i) *Integration of TCGA tissue expression and methylation data*. The process starts with creating a subset of a cohort consisting of tissue samples with both expression and methylation data. For each T cell marker gene, Pearson correlation (R) between expression level and methylation of each CpG site is calculated for a select group of genes crucial to T cell-related pathways (*CD3D, CD3E, CD3G, CD247, CTLA4, PDCD1, CD274, GZMA, PRF1*), through a previously described approach (Bacolod et al., 2015). (ii) *Assessment of the effects of CpG methylation on histone constitution at the promoter region*. Another data repository which is very useful for cancer research is ENCODE (Encyclopedia of DNA Elements) (https://genome.ucsc.edu/ENCODE/) (2004), a project which aims to build a comprehensive list of functional elements in the human genome, and further our understanding of gene regulations. In ENCODE, numerous experimental tools, such as ChIP Seq technology, were employed to examine how proteins (e.g., transcription factors, histones) recognize and bind to genomic sequences such as promoter regions. CpG methylation data (Illumina 450K) for certain cell lines were extracted from datasets GSE40699 and GSE68379 (see above). Through the UCSC Genome Browser, we are able to visualize the association between level of CpG methylation and intensity of bound modified histones at the promoter region. The latter information was contributed by the B.E. Bernstein's lab (Broad Inst.) using ChIP Sequencing (Bernstein et al., 2005). (iii) *Assessment of the effects of CpG methylation on DNase hypersensitivity at the promoter region*. It is also possible to examine the association between promoter methylation and the accessibility of the chromatin by overlapping CpG methylation data and hypersensitivity to DNase I, which is generated by sequencing of DNase I-treated genomic DNA, and also accessible through the UCSC Genome Browser (Boyle et al., 2008).

3. Results

3.1 Gene set enrichment analyses (GSEA) indicate differences in the extent of relative activation of T cells and other immune signaling pathways, in different types of cancer

As anticipated, the Biocarta gene sets for cell cycle and proliferation, as well as pathways indicative of fast dividing cells, registered the highest ES values for each cohort (Fig. 2). These include the gene sets: G1 (*Cell Cycle: G1/S Check Point*), G2 (*Cell Cycle: G2/M Check*point), MCM (*CDK Regulation of DNA Replication*), RB (*RB Tumor Suppressor/Checkpoint Signaling in Response to DNA Damage*), P27 (*Regulation of p27 Phosphorylation During Cell Cycle Progression*), P53 (*p53 Signaling Pathway*), CELLCYCLE (*Cyclins and Cell Cycle Regulation*), ATM (*ATM Signaling Pathway*), ARF (*Tumor Suppressor ARF Inhibits Ribosomal Biogenesis*), ATRBRCA (*Role of BRCA1, BRCA2 and ATR in Cancer Susceptibility*). Indeed, the gene sets G1, G2, P27, MCM, CELLCYCLE pathways are among the top 30 most activated pathways in 16 of 18 TCGA cohorts. The same analysis indicate that no cohort is entirely free of upregulation of immune involvement, given that the top 30 activated pathways in every cohort are immune-related pathways (Fig. 3). In descending order, the proportion of immune related pathways (among the top 30) are as follows: SKCM (25/30), KIRC (24/30), KIRP (18/30), HNSC (14/30), ESCA (12/30), PRAD (12/30), BRCA (11/30), KICH (11/30), UCEC (9/30), THCA (7/30), STAD (7/30), COAD (4/30), READ (4/30), LUAD (4/30), LUSC (4/30), LIHC (3/30), BLCA (3/30), and CHOL (2/30).

The pathways related to immune infiltration and activation can roughly be divided into those related to: (a) infiltration by helper T lymphocytes, (b) infiltration by cytotoxic T lymphocytes, (c) T cell activation, (d) natural killer cells, (d) B cell infiltration, (e) cytokines and inflammation. Many of these pathways have overlapping components, and not mutually

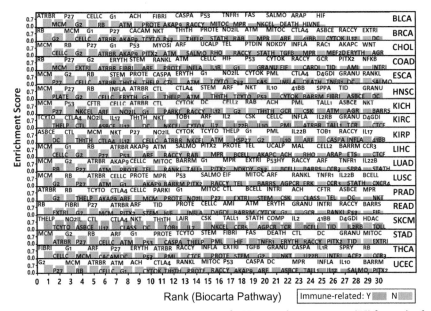

Fig. 2 The top 30 Biocarta pathways in terms of GSEA enrichment scores (ES) for each of the 18 TCGA cohorts (with substantial, i.e., at least 9 reference tissues). Except for SKCM, ES refers to statistical comparison between primary tumors and normal. For SKCM, the comparison is between metastasis and primary tumors. The Biocarta Pathways represented in the bar graphs are indicated by capitalized abbreviations (e.g., "RB" refers to *"RB Tumor Suppressor/Checkpoint Signaling in Response to DNA Damage CTLA4"*). Whether a Biocarta pathways is immune-related (orange) or not (aquamarine) is determined through manual inspection of each pathway by the authors (MDB, RDC). Due to limited space, for certain Bicoarta gene sets, the IDs displayed are the truncated version of the original, such as: AKAP9 (original ID is AKAP95), ASBCE (ASBCELL), ATRBR (ATRBRCA), BARRM (BARR MAPK), BARRS (BARRESTIN SRC), BCELL (BCELLSURVIVAL), CASPA (CASPASE), CELL2 (CELL2CELL), CELLC (CELLCYCLE), CLASS (CLASSIC), CYTOK (CYTOKINE), EXTRI (EXTRINSIC), FIBRI (FIBRINOLYSIS), GRANU (GRANULOCYTES), INFLAM (INFLA), MITOC (MITOCHONDRIA), MYOSI (MYOSIN), NKCEL (NKCELLS), NO2IL (NO2IL12), PARKI (PARKIN), PLATE (PLATELETAPP), PROTE (PROTEASOME), PTDIN (PTDINS), RACCY (RACCYCD), SALMO (SALMONELLA), STATH (STATHMIN), TCYTO (TCYTOTOXIC), TH1TH (TH1TH2), THELP (THELPER), UCALP (UCALPAIN).

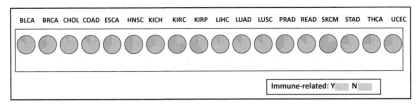

Fig. 3 Fraction of immune-related pathways in Fig. 2, for each cohort.

exclusive from each other. Among the inter-connected gene sets and pathways representing infiltrating T cells are: THELPER (*T Helper Cell Surface Molecules*), TH1TH2 (*Th1/Th2 Differentiation*), DC (*Dendritic Cells in Regulating Th1 and Th2 Development*), IL12 (*IL12- and Stat4-Dependent Signaling Pathway in Th1 Development*), and NKT (*Selective Expression of Chemokine Receptors During T-cell Polarization*). The aforementioned gene sets, reflective of a highly effective priming of tumor-specific T cells, describe how the interaction of an antigen presenting cell with precursor T helper cell eventually leads to the latter's differentiation into two possible types: the IL12-induced Th1 and IL4-induced Th2 helper T cells. In the CTLA4 pathway (*The Co-Stimulatory Signal During T-cell Activation*) (Fig. 4A), the binding of T cell's CD28 surface protein to cancer cell's CD80 (or CD86) counterpart can lead to activation of PI3K, LCK, and GRB signaling, toward T cell proliferation. This process can be halted when T cell's CTLA4 is instead bound to APC's CD80. The DC pathway (Fig. 4B) points to the crucial role dendritic cells play in Th1/Th2 differentiation, with myeloid (pDC1) and lymphoid (pDC2) dendritic cells and respectively driving Th1 and Th2 cell differentiation. The NKT pathway describes how Th1 and Th2 cells acquire differences in chemokine receptor expression upon polarization of naïve T cells. The infiltration by cytotoxic T cells is represented by the gene sets CTL (*Cytotoxic T cell-Mediated Immune Response Against Target Cells*) and TCYTO (*T Cytotoxic Cell Surface Molecules*). Both gene sets describe how cytotoxic T cells induce the apoptosis of an antigen presenting cancer cell through either the Fas or Caspase pathways. These gene sets are also characterized by upregulation of cytotoxic T cell-specific membrane marker protein CD8, FASLG (Fas Ligand), and Granzyme B (protease secreted by cytotoxic T cells). Other gene sets focus on the mechanisms controlling the activation of T cells toward the elimination of cancer cells. In the CSK pathway (*Activation of Csk by cAMP-Dependent Protein Kinase Inhibits Signaling Through the T Cell Receptor*), the interaction of T cell receptor with antigen-presenting cells can signal LCK to phosphorylate ZAP70, which will then activate the MAP kinase pathway toward T cell activation. The protein CSK can act as a modulator of this process by phosphorylating (and deactivating) LCK. T cell activation (upon the interaction between APC and T cells) can also be relayed through the TNF receptor 4-1BB present on the T cell surface (41BB; *The 4-1BB-Dependent Immune Response*). Unlike T cells, natural killer (NK) cells can recognize their target cells (such as tumor cells) without antigen presentation through MHC. As described in the pathway NKCELLS (*Ras-Independent Pathway in NK Cell-Mediated Cytotoxicity*),

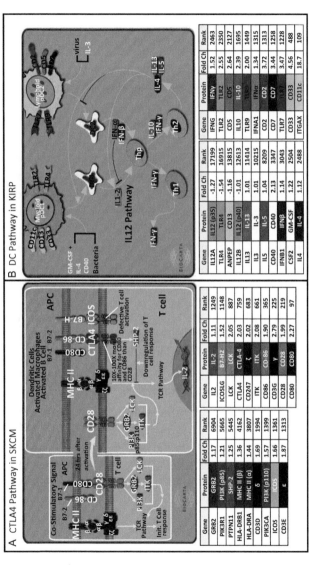

Fig. 4 The upregulation of the CTLA-4 and DC pathways in SKCM (A) and KIRP (B), respectively, demonstrated through GSEA. (A) Shown is a cartoon of the Biocarta CTLA-4 pathway ("*The Co-Stimulatory Signal During T-cell Activation*"). Relative to primary tumors, the metastasis samples in the SKCM cohort exhibited elevated expression of most of the genes coding for proteins involved in the pathway (see inset table). (B) Similarly, the Biocarta DC pathway ("*Dendritic Cells in Regulating Th1 and Th2 Development*") is more active in the KIRP primary tumors relative to solid normal. As shown in the inset table, numerous genes involved in this pathway (including surface proteins in dendritic cells) exhibited elevated expression (as fold-change) in primary tumors relative to normal.

NK cell's recognition of tumor cell results in the activation of glycoprotein receptors (such as the NKG2D receptor) on its surface, followed by a signaling cascade and release of perforins to penetrate and promote apoptosis of the target cancer cells. In the NOTIL pathway (*NO2-Dependent IL12 Pathway in NK Cells*), a macrophage may secrete IL12, which binds to IL12R on NK cells. This will activate the JAK/STAT signaling pathway, leading to transcriptional activation of IFN Gamma which then induces the differentiation of Thp to Th1 cells. There are also gene sets indicative of cytokines and inflammations: CYTOK (*Cytokine Network*) and INFLA (*Cytokines and Inflammatory Response*). As expected, two gene sets encompassing mostly cytokines (CYTOK or *Cytokine Network*, and INFLA or *Cytokines and Inflammatory Response*) exhibit high ES values in various cancers (mostly those which also registered upregulation in pathways related to T cells, NK cells, and DC cells, in which many of these cytokines are components). Leukocyte infiltration around tumor cells may explain the high ES values for gene set GRANULOCYTES (*Adhesion and Diapedesis of Granulocytes*). Surface markers known to be expressed in neutrophils include CD11B/ITGAM (which along with CD18 forms MAC-1, which is crucial to the leukocyte's adhesion to the target cell), as well as IL8 and CSF3 (both important in regulating this process).

3.2 Renal, skin, head and neck, and esophageal cancers exhibit the most highly enriched pathways related to activation and proliferation of helper and cytotoxic T cells

Comparative GSEA between primary tumors and solid normal tissues for 17 of the 18 cohorts (for SKCM, the comparison is between metastasis and primary tumors) reveal that certain cancer types activate immune processes more prominently relative to others. Indeed, immune-related pathways (particularly those related to T cell infiltration and activation) represent higher proportion among the top 30 differentially activated pathways in kidney, skin, head and neck, breast, endometrial, esophageal, and prostate cancers.

3.2.1 Kidney cancers
Common to these three kidney cohorts (KIRC, KIRP, KICH) are high ES values for Biocarta pathways/genes associated with helper and cytotoxic T cells (TH1TH2, DC, IL12, NKT). In KIRC and KIRP cohorts, the activation of T cells is manifested by the high ES values for CTLA4 and TOB1 (*Role of Tob in T-cell Activation*) pathways. The CSK pathway also registered

high ES values for both KIRC and KICH. For the three cohorts, the infiltration by Natural Killer cells is highly evident given the high ES values for NKCELLS and NO2IL12, while B cell infiltration is likely indicated by high ES values for the gene set ASBCELL (*Antigen-Dependent B Cell Activation*). KIRC exhibits high ES values for GRANULOCYTES and IL17 (*IL17 Signaling Pathway*), which are indicators of granulocyte infiltration. The upregulation of cytokines is likely for each kidney cohort given the high ES values for the gene set CYTOKINE. The upregulation of T helper cell pathways can be explained by upregulation of IL12R (consisting of IL12RB1 and IL12RB2, which are present in both Th1 and Th2 lymphocytes), IL4R, and STAT4 (the mediator of IL12's role in Th1 maturation). Results also indicate that both Th1 (such as the pDC1 surface marker CD33 and the chemokine receptors CXCR3 and CCR1) and Th2 (such as the pDC2 surface marker CD5 and TLR7, and the chemokine receptor CCR) markers are highly upregulated in kidney as well as esophageal, skin, endometrial, head and neck, breast, thyroid, stomach, and prostate cancers. The likely infiltration by NK cells in kidney (as well as esophageal, head and neck, breast, endometrial, and skin) cancers can be explained by elevated expression of NK-specific receptors such as KLRC1 (Ly49), KLRC2, and KLRD1 (CD94), as well as LAT, which are involved in the NOTIL12 pathway.

3.2.2 Melanoma
Similar to the renal cancers, metastatic melanoma (relative to primary tumors) exhibit high ES values for pathways indicative of infiltration by helper T lymphocytes (THELPER, TH1TH2, DC IL12, NKT, TCR), cytotoxic T lymphocytes (TCYTOTOXIC, CTL), natural killer cells (NKCELLS, NO2IL12), B cells (BCELLSURVIVAL, ASBCELL, TALL1), macrophages (CCR5), and T cell activation (CTLA4, 41BB, CSK, IL7, IL2, IL2RB, TOB1).

3.2.3 Head and neck cancer
Relative to normal, primary HNSC tumors exhibit upregulation of pathways related to infiltration (TH1TH2, THELPER, DC, NKT, TCYTOTOXIC, CTL) and activation (41BB, CTLA4) of T lymphocytes. Also evident is infiltration of granulocytes (GRANULOCYTES). Overall, these results are consistent with the observed upregulation of markers specific for CD8 T cells, Cytotoxic T cells, T cells, DC, and Neutrophils.

3.2.4 Breast cancer

Comparative (primary tumors vs. normal) analysis indicates upregulation of pathways associated with helper T lymphocytes (TH1TH2, THELPER, DC, IL12, NKT), cytotoxic T lymphocytes (TCYTOTOXIC), natural killer cells (NO2IL12), B lymphocytes (ASBCELL), as well as T cell activation (CTLA4, 41BB).

3.2.5 Endometrial cancer

Evident in endometrial cancer are the high ES values for pathways related to T helper lymphocyte proliferation (TH1TH2, DC, IL12) and activation (CTLA4), as well as B cell infiltration (ASBCELL, TALL1). These observations are largely consistent with upregulation of genes that are markers for CD8 T, B cells, Cyto T cells, and DC.

3.2.6 Esophageal cancer

The primary tumors of esophageal cancer exhibit upregulation of pathways associated with helper T lymphocytes (TH1TH2, THELPER, DC), cytotoxic T lymphocytes (CTL, TCYTOTOXIC), T cell activation (CTLA4), natural killer cells (NO2IL12), and granulocytes (GRANULOCYTES).

3.2.7 Prostate cancer

For prostate cancer, bioinformatic analysis points to the upregulation of pathways related to helper (DC, NKT, THELPER) and cytotoxic (CTL, TCYTOTOXIC) T cells, the T cell activation processes (CSK, CTLA4), infiltration by natural killer cells (NO2IL12), and B cell proliferation (ASBCELL, BCELLSURVIVAL).

3.2.8 Thyroid cancer

The infiltration by helper T lymphocytes in thyroid cancer is marked by high ES values for the gene sets DC and NKT. Likewise, the high ES value for the gene set GRANULOCYTES is indicative of presence of granulocytes in thyroid cancer PTs.

3.3 Relative to normal tissues, T cell-related pathways are not significantly enriched in the primary tumors of lung, colorectal, liver and bile duct cancers

In contrast to the cancer types described above, the upregulation of T lymphocyte-related processes is not as highly evident in colorectal, lung, liver/bile duct, and bladder cancers. However, it does not necessarily mean that these tumors are devoid of T cell infiltration. It is possible that the surrounding normal tissues are themselves infiltrated with immune cells.

3.3.1 Lung cancer
Although LUAD and LUSC exhibit the highest average expression of T cell markers among all the primary tumors, these markers are even more highly expressed in normal lung tissues. This explains the relatively low ES values for T cell-related pathways in these two cohorts. Nevertheless, LUAD and LUSC cohorts registered relatively high ES values for B CELL SURVIVAL (indicative of B cells) and CCR3 (indicative of granulocyte) pathways.

3.3.2 Colorectal and stomach cancers
Just as noted in lung cancer, the upregulation of T cell-related pathways is not observed in COAD and READ cohorts. Both cohorts also exhibit relatively high ES for GRANULOCYTE. Nevertheless, the histologically related STAD registers relatively high ES for pathways indicative of helper T (DC) and cytotoxic T (CTL, TCYTOTOXIC) lymphocyte infiltration.

3.3.3 Liver and bile duct cancers
T cell-related processes are not upregulated in liver (LIHC) and bile duct (CHOL) primary tumors. LIHC (but not CHOL) displays high ES values for B CELL SURVIVAL and CCR3 suggesting the possibility of B cell and granulocyte infiltration, respectively.

3.3.4 Bladder cancer
Just like in lung, colorectal and liver, the upregulation of T cell infiltration-related pathways is not evident in bladder cancer. However, there is likely infiltration of natural killer cells (NKCELLS) in this particular cancer type.

3.4 For certain cancer types, immune infiltration/activation correlates with improved prognosis

The prognostic values of immune activation can be assessed bioinformatically by relating the expression values of immune-related genes (individually or as a group of genes involved in particular pathways) to clinical follow-up records. Fortunately, TCGA datasets are accompanied by patient survival records; thus we were able to apply Cox's proportional hazards model to calculate statistical measures of prognosis (i.e., hazard ratio, parametric P value) when the primary tumors are grouped into either low (below median) or high (above median) expression group. Gene Set Analysis

(GSA) (Efron & Tibshirani, 2007) was also applied to assess the survival risk associated with each of the 187 Biocarta pathways (those with at least 12 member genes). Since no control tissues were needed, survival analyses were conducted for all the 32 TCGA cohorts. In this test, a negative P value (for a Biocarta pathway) can be interpreted as the activation of that particular pathway being associated with good prognosis. On the other hand, a positive P value means activation of that pathway correlating with poor outcome. The cohorts having Biocarta pathways considered immune-related and whose activations are associated with good prognosis are: SKCM (8 pathways), HNSC (6), CESC (5), SARC (4), BLCA (2), MESO (2), STAD (2), BRCA (1), LGG (1), LIHC (1), OV (1), KIRC (1), and LUAD (1) (Table 1). However, the aforementioned cohorts do not necessarily have similar immune-related, good prognosis-correlated pathways. The pathways related to T lymphocyte activation (such as TCR, CTL, TH1TH2, CSK, IL12, as well as CTLA4) have positive prognostic values in CESC, HNSC, and SKCM. NKCELLS appear positively prognostic in BLCA and SKCM,

Table 1 Immune-related biocarta pathways with significant prognostic values in solid tumors as identified using the "Survival Gene Set Analysis" functionality of BRB ArrayTools. The identified pathways exhibit $P < 0.01$ in the Efron-Tibshirani GSA test. In addition, the median LS (least squares) and KS (Kolmogorov-Smirnov) permutation P values for these pathways are 0.0017 and 0.0062 respectively.

Cohort	Biocarta pathway	Number of genes	Efron-Tibshirani's GSA test P-value	Sign, Efron-Tibshirani's GSA test P-value	Pathway description
ACC	BCELLSURVIVAL	15	0.005	(+)	B Cell Survival Pathway
BLCA	NKCELLS	20	0.005	(−)	Ras-Independent pathway in NK cell-mediated cytotoxicity
BLCA	PCAF	13	<0.005	(−)	The information-processing pathway at the IFN-beta enhancer
BRCA	PCAF	13	<0.005	(−)	The information-processing pathway at the IFN-beta enhancer
CESC	CSK	19	0.005	(−)	Activation of Csk by cAMP-dependent Protein Kinase Inhibits Signaling through the T Cell Receptor

Table 1 Immune-related biocarta pathways with significant prognostic values in solid tumors as identified using the "Survival Gene Set Analysis" functionality of BRB ArrayTools. The identified pathways exhibit $P < 0.01$ in the Efron-Tibshirani GSA test. In addition, the median LS (least squares) and KS (Kolmogorov-Smirnov) permutation P values for these pathways are 0.0017 and 0.0062 respectively.—cont'd

Cohort	Biocarta pathway	Number of genes	Efron-Tibshirani's GSA test P-value	Sign, Efron-Tibshirani's GSA test P-value	Pathway description
CESC	CTLA4	19	0.005	(−)	The Co-Stimulatory Signal During T-cell Activation
CESC	IL12	19	0.005	(−)	IL12 and Stat4 Dependent Signaling Pathway in Th1 Development
CESC	NO2IL12	15	<0.005	(−)	NO2-dependent IL 12 Pathway in NK cells
CESC	TCR	44	0.005	(−)	T Cell Receptor Signaling Pathway
ESCA	CYTOKINE	19	<0.005	(+)	Cytokine Network
ESCA	DC	17	0.005	(+)	Dendritic cells in regulating TH1 and TH2 Development
ESCA	IL17	15	<0.005	(+)	IL 17 Signaling Pathway
ESCA	INFLAM	25	0.005	(+)	Cytokines and Inflammatory Response
HNSC	CSK	19	0.005	(−)	Activation of Csk by cAMP-dependent Protein Kinase Inhibits Signaling through the T Cell Receptor
HNSC	FCER1	39	0.01	(−)	Fc Epsilon Receptor I Signaling in Mast Cells
HNSC	IL17	15	0.005	(−)	IL 17 Signaling Pathway
HNSC	IL2	22	0.005	(−)	IL 2 signaling pathway
HNSC	NO2IL12	15	0.01	(−)	NO2-dependent IL 12 Pathway in NK cells
HNSC	TCR	44	<0.005	(−)	T Cell Receptor Signaling Pathway

Continued

Table 1 Immune-related biocarta pathways with significant prognostic values in solid tumors as identified using the "Survival Gene Set Analysis" functionality of BRB ArrayTools. The identified pathways exhibit $P < 0.01$ in the Efron-Tibshirani GSA test. In addition, the median LS (least squares) and KS (Kolmogorov-Smirnov) permutation P values for these pathways are 0.0017 and 0.0062 respectively.—cont'd

Cohort	Biocarta pathway	Number of genes	Efron-Tibshirani's GSA test P-value	Sign, Efron-Tibshirani's GSA test P-value	Pathway description
KIRC	GSK3	20	0.005	(−)	Inactivation of Gsk3 by AKT causes accumulation of b-catenin in Alveolar Macrophages
LGG	CDMAC	16	0.005	(−)	Cadmium induces DNA synthesis and proliferation in macrophages
LIHC	IL17	15	0.01	(−)	IL 17 Signaling Pathway
LUAD	BCR	33	<0.005	(−)	BCR Signaling Pathway
LUSC	CLASSIC	13	0.01	(+)	Classical Complement Pathway
LUSC	COMP	18	0.01	(+)	Complement Pathway
MESO	CLASSIC	13	0.005	(−)	Classical Complement Pathway
MESO	COMP	18	0.005	(−)	Complement Pathway
OV	IL17	15	0.005	(−)	IL 17 Signaling Pathway
OV	GSK3	20	0.005	(+)	Inactivation of Gsk3 by AKT causes accumulation of b-catenin in Alveolar Macrophages
READ	FMLP	34	0.005	(+)	fMLP induced chemokine gene expression in HMC-1 cells
SARC	CD40	15	0.005	(−)	CD40L Signaling Pathway

Table 1 Immune-related biocarta pathways with significant prognostic values in solid tumors as identified using the "Survival Gene Set Analysis" functionality of BRB ArrayTools. The identified pathways exhibit $P < 0.01$ in the Efron-Tibshirani GSA test. In addition, the median LS (least squares) and KS (Kolmogorov-Smirnov) permutation P values for these pathways are 0.0017 and 0.0062 respectively.—cont'd

Cohort	Biocarta pathway	Number of genes	Efron-Tibshirani's GSA test P-value	Sign, Efron-Tibshirani's GSA test P-value	Pathway description
SARC	GRANULOCYTES	15	0.01	(−)	Adhesion and Diapedesis of Granulocytes
SARC	IL10	13	0.005	(−)	IL-10 Anti-inflammatory Signaling Pathway
SARC	PCAF	13	<0.005	(−)	The information-processing pathway at the IFN-beta enhancer
SKCM	CSK	19	0.01	(−)	Activation of Csk by cAMP-dependent Protein Kinase Inhibits Signaling through the T Cell Receptor
SKCM	CTL	13	<0.005	(−)	CTL mediated immune response against target cells
SKCM	CTLA4	19	0.005	(−)	The Co-Stimulatory Signal During T-cell Activation
SKCM	IL12	19	0.01	(−)	IL12 and Stat4 Dependent Signaling Pathway in Th1 Development
SKCM	INFLAM	26	0.005	(−)	Cytokines and Inflammatory Response
SKCM	NKCELLS	20	<0.005	(−)	Ras-Independent pathway in NK cell-mediated cytotoxicity
SKCM	NO2IL12	15	0.005	(−)	NO2-dependent IL 12 Pathway in NK cells
SKCM	TH1TH2	20	<0.005	(−)	Th1/Th2 Differentiation

Continued

Table 1 Immune-related biocarta pathways with significant prognostic values in solid tumors as identified using the "Survival Gene Set Analysis" functionality of BRB ArrayTools. The identified pathways exhibit $P < 0.01$ in the Efron-Tibshirani GSA test. In addition, the median LS (least squares) and KS (Kolmogorov-Smirnov) permutation P values for these pathways are 0.0017 and 0.0062 respectively.—cont'd

Cohort	Biocarta pathway	Number of genes	Efron-Tibshirani's GSA test P-value	Sign, Efron-Tibshirani's GSA test P-value	Pathway description
STAD	TID	18	<0.005	(−)	Chaperones modulate interferon Signaling Pathway
STAD	TNFR2	17	<0.005	(−)	TNFR2 Signaling Pathway
TGCT	CD40	15	0.01	(+)	CD40L Signaling Pathway
TGCT	GRANULOCYTES	15	0.01	(+)	Adhesion and Diapedesis of Granulocytes
TGCT	LYM	14	<0.005	(+)	Adhesion and Diapedesis of Lymphocytes
THYM	VIP	26	<0.005	(−)	Neuropeptides VIP and PACAP inhibit the apoptosis of activated T cells

while activation of PCAF (*The information-processing pathway at the IFN-beta enhancer*) is associated with higher survival rate in BLCA, BRCA, and SARC. One cohort in particular (OV) simultaneously had both good prognosis- and bad prognosis-correlated immune pathways: IL17 (*IL17 Signaling*) and GSK3 (*Inactivation of Gsk3 by AKT Causes Accumulation of β-catenin in Alveolar Macrophages*) activation associated with favorable and unfavorable prognosis, respectively. A few of these immune pathways that reached prognostic significance can have a good-prognostic value in one cohort while having a bad-prognostic value in another: IL17 is good for HNSC, LIHC and OV, while bad for ESCA; GSK3 is good for KIRC while bad for OV; and CLASSIC is good for MESO while bad for LUSC. Many pathways related to immune cell activation have overlapping component genes. A closer look at how these genes (not the pathways) individually influence prognosis is illustrated in Fig. 5. Shown is a heatmap of parametric P values for each of the genes belonging to CTLA4, THELPER,

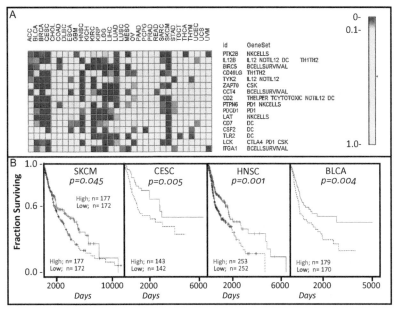

Fig. 5 (A) Out of the 126 genes which are components of the Biocarta (BCELLSURVIVAL, CSK, CTLA4, DC, GRANULOCYTES, IL12, NKCELLS, NOTIL12, PD1, TCYTOTOXIC, TH1TH2, THELPERnnn) and Reactome (PD1) Immune-related pathways, the 16 genes depicted in the heatmap exhibit parametric P values 0.05 (or lower) in at least 9 cohorts. This indicates that many of the genes listed (including PDCD1) may have prognostic values. (B) As shown in these Kaplan-Meier plots, higher expression of the gene LAT (Linker for Activation of T-Cell) is associated with better outcome for patients with melanoma (SKCM), cervical (CESC), head and neck (HNSC), and bladder (BLCA) cancers. In these graphs, the patients divided across median of LAT expression.

TCYTOTOXIC, CSK, IL12, NOTIL12, DC, and NKCELLS pathways, for the different TCGA cohorts. Of particular interest is a cluster of low P values (Fig. 5A) which turn out to include the P values for T lymphocyte activation checkpoint genes (CTLA-4, PD-1), components of T cell receptor (CD3D, CD3G, CD3E), CD80, CD247, CD8A, ICOS, ITK, ZAP70, and LAT across a group of cohorts including BLCA, BRCA, CESC, HNSC, LGG, LIHC, OV, SARC, LUAD, and SKCM. Most of these low P values correspond to $HR < 1$, meaning higher expression of these genes points to better clinical outcome (longer survival), as exemplified in the gene LAT (Fig. 5B). CTLA-4 is positively prognostic for SKCM, as well as for CESC and HNSC. PD-1 (PDCD1) is also positively prognostic for SKCM, and BLCA, BRCA, CESC, HNSC and SARC, but negatively prognostic for LGG.

3.5 Identification of CpG promoter sites most highly associated with expression of important T lymphocyte marker genes

All of the 32 TCGA cancer cohorts are represented by both mRNA transcriptional and methylation (Illumina Infinium Human Methylation 450K) datasets. The Illumina Methylation array is designed to interrogate >480,000 CpG sites in the entire human genome (Sandoval et al., 2011). The integration of the methylation and transcriptional datasets makes it possible to examine the likely association between the expression levels of genes most prominently associated with T cell infiltration and activation. For this particular analysis, we excluded OV (Ovarian serous cystadenocarcinoma) cohort since the available methylation dataset was generated by the much less dense Illumina 27K array (interrogating only 27,578 CpG sites). We focused on predicting the expression/methylation correlations for genes comprising T cell receptor (*CD3D, CD3E, CD3G, CD247*), the checkpoint inhibitor target genes (*CTLA4, PDCD1*), and genes indicative of T cell's cytolytic activity (*GZMA, PRF1*). These genes have varying numbers of interrogated CpG sites, mostly within the vicinity of the transcriptional start site (TSS) (see Figs. 6 and 7). Through JMP's multivariate analysis functionality, the Pearson R (or R^2) values between gene expression and the methylation level of each interrogated local CpG site are calculated for each of the 31 cancer cohorts, and plotted in heatmaps (Figs. 6 and 7). The heatmap spectrum ranges from green (relatively high negative R values) to gray to red (relatively high positive R values). A green square is therefore indicative of CpG methylation negatively correlating with the expression level of a given gene. Indeed, there is an apparent cluster of green squares at the promoter regions of the TCR genes (*CD3D, CD3E, CD3G, CD247*) for most of the 31 cancer types. These observations suggest that these CpG sites likely factor in the transactivation of these genes. As we look more closely, the methylation at cg07728874 (located at the first exon) is most negatively correlated with the expression of CD3D (as indicated by median R value of −0.4424 for the 31 cohorts). Nevertheless, there is a noticeable variation across cancer types. While the cg07728874 R values for THYM, THCA, and KIRP are −0.9256, −0.6922, and −0.645, respectively (see Fig. 8 for the plots), the values for GBM, PCPG, and LGG are 0.1253, 0.0596, and 0.1253, respectively. Similar trends are observed for cg24612198 (5′ UTR of CD3E), cg15880738 (located in 5′ UTR of CD3G), and cg07786657 (within intron 1, but <100-bp from the transcription start site in CD247 locus), which register the lowest median R values

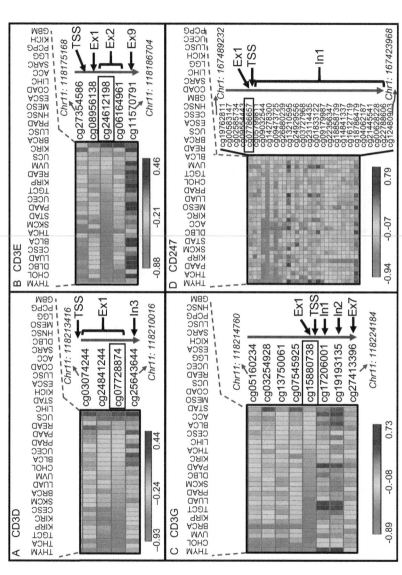

Fig. 6 Heatmaps representing the Pearson R values for CD3D (A), CD3E (B), CD3G (C), and CD247 (D) expression vs. methylation at local CpG sites (indicated by its CpG ID, and genomic coordinate or GC). Also indicated are the relative location of the transcriptional start site (TS), and whether the CpG site is part of a particular Exon (Ex) or Intron (In). The columns are arranged in order of increasing (i.e., more positive) R for the CpG site with the lowest median R across tumor types (cg07728874 for CD3D, cg24612198 for CD3E, cg15880738 for CD3G, and cg07786657 for CD247).

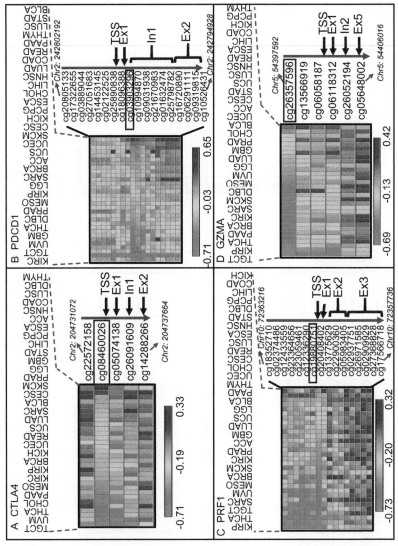

Fig. 7 Heatmaps representing the Pearson R values for CTLA4 (A), PDCD1 (B), PRF1 (C), and GZMA (D) expression vs. methylation at local CpG sites (indicated by its CpG ID, and genomic coordinate or GC). Also indicated are the relative location of the transcriptional start site (TS), and whether the CpG site is part of a particular Exon (Ex) or Intron (In). The columns are arranged in order of increasing (i.e., more positive) R for the CpG site with the lowest median R across tumor types.

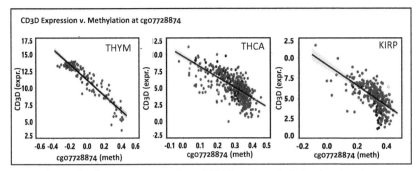

Fig. 8 Relationship between expression of CD3D and methylation level at the promoter region CpG site cg07728874 in THYM, THCA, and KIRP cohorts.

(calculated for the 31 cohorts) within the respective gene locus. Nonetheless, these R values may vary across cohorts. For example, the average R for the three aforementioned CpG sites are −0.8819 for THM, −0.6720 for THCA, −0.7681 for CHOL, −0.1043 for GBM, −0.1545 for PCPG, and −0.3154 for LGG. This suggests that methylation-driven activation for these T cell receptor genes is much more pronounced in thymoma, thyroid cancer, and bile duct cancer, compared to brain and neuroendocrine tumors. Similar analyses were conducted to infer the CpG sites most likely to influence the transcription of the other immune-marker genes. These are: cg03903296 (median R = −0.2159, for PDCD1), cg08460026 (median R = −0.3101, for CTLA4), cg26357596 (median R = −0.4474, for GZMA), and cg19880751 (median R = −0.4338, for PRF1). Just like in T cell receptor genes, the effects of CpG methylation on the transcription of these genes appear more likely in certain cohorts. The R values for both cg08460026 (CTLA4) and cg03903296 (PDCD1) are more negative in KIRC (−0.6698, −0.706), TGCT (−0.6698, −0.5774), UVM (−0.6438, −0.5348), but are not very significant in THYM (0.1791, 0.1761) and LUSC (−0.0973, 0.189).

3.6 Promoter CpG methylation likely influences the binding of modified histones, and thereby the transcription of immune-related genes

It would also be of interest to investigate how these CpG sites possibly affect the constitution of chromatin around their respective promoter regions. This is accomplished through analyses of ChIP Seq data for various modified histones, publicly available through the ENCODE project and the UCSC Genome Browser. Take for example the CpG sites cg03074244,

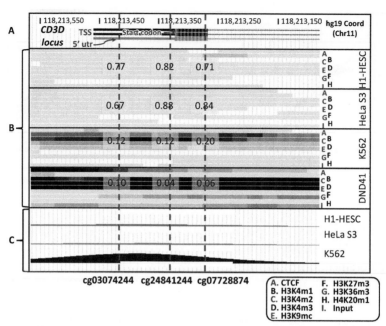

Fig. 9 (A) The promoter region of the CD3D gene. (B) The "affinity" of modified histones and CTCF toward the CD3D promoter region of the H1-hESC, HeLa S3, K562, and DND41 cell lines. The figure indicates positive association between intensity of modified histones (such as H3K4m1, H3K4m2, H3K4m3, H2K27m3) and the methylation at 3 CpG sites (shown are the methylation β values). (C) Cell line such as K562 also happens to be more hypersensitive to DNase I (compared to H1-HESC and HeLa S3).

cg24841244, and cg07728874 (discussed earlier), which based on our calculations are most likely associated with the transcription of CD3D. According to GSE40699 and GSE68379 datasets, these CpG sites are highly methylated in the human embryonic stem cell line H1-HESC (β: 0.77, 0.82, 0.71, respectively) and the cervical cancer cell line HeLa S3 (β: 0.67, 0.88, 0.84), but lowly methylated in the chronic myelogenous leukemia cell line K562 (β: 0.12, 0.12, 0.20), and the childhood T acute lymphoblastic leukemia cell line DND41 (β = 0.10, 0.04, 0.06) (Fig. 9A and B). It is also clear that for the cell lines DND41 and K562, the CD3D promoter region encompassing the three aforementioned CpG sites also happens to exhibit high affinity (darker bands) for Histone H3 mono-, di-, or trimethylated at K4 (H3K4me1, H3K4me2, H3K4me3), and acetylated at K9 (H3K9ac). In contrast, the same bands seem to be absent in H1-hESC and HeLa S3. Those histone modifications are associated with a more active transcription (Campbell & Turner, 2013; Ernst et al., 2011; Ghirlando et al., 2012), suggesting that CD3D is more actively transcribed in the heme-derived (DND41, K562) compared to epithelial (HeLa S3) and embryonic (H1-HESC) cell lines. Another indication that hypomethylation at these CpG sites is associated with elevated rates of transcription is the fact that DNase I hypersensitivity (as indicated by density of DNA fragments resulting from DNase I digestion) at this particular region is more intense in K562 compared to HeLa S3 and H1-HESC (Note that DNase I hypersensitivity data for DND41 are not available) (Fig. 9C).

4. Discussion

The primary bioinformatic approach (i.e., identification of pathways activated in tumor relative to normal samples, or one sample subset over another, using Gene Set enrichment Analysis or GSEA) was reliably employed in a wide-range of cancer-related studies. Previously, Bacolod and colleagues utilized GSEA to identify the pathways activated in *MDA-9/Syntenin*-overexpressing vs. *MDA-9/Syntenin*-under-expressing cancers (Bacolod et al., 2015), as well as high risk vs. low risk (Bacolod et al., 2016) gliomas. Similar approaches were also applied to cancer progression in bladder cancer (Chen et al., 2016), lung cancer (Cai & Jiang, 2014), and triple negative breast cancer (Narrandes, Huang, Murphy, & Xu, 2018). In this current report, the comparison of primary tumor vs. normal samples in 17 cancer types (the lone exception being melanoma, in which the comparison was between metastatic and primary tumors) identified the activation of various pathways and processes associated with cell division and proliferation that are readily associated with cancer progression (e.g., *p53 and p27 signaling pathways, cyclin and cell cycle regulation*). The elevated tumor expression of component genes renders these pathways with high ES values, which can be interpreted as indicative of activation. For example, the pathway G2 (*Cell Cycle: G2/M Checkpoint*) is activated in most cohorts due to upregulation of target genes *CHEK2, CDK1, CD25B, CHEK1,* and *PLK1*. Moreover, the bioinformatic identification of cancer progression pathways supports the validity of similarly identified immune-related processes, which appear to be most highly enriched in renal, skin, head and neck, esophageal, breast, cervical, endometrial, prostate, and thyroid cancers. These immune-related Biocarta gene groups (depicting pathways, processes, or cell types) include: TH1TH2 (*Th1/Th2 Differentiation*), DC (*Dendritic Cells in Regulating Th1 and Th2 Development*), IL12 (*IL12 and Stat4 Dependent Signaling Pathway in Th1 Development*), NKT (*Selective Expression of*

Chemokine Receptors During T-Cell Polarization), CTL (*CTL Mediated Immune Response Against Target Cells*), CTLA4 (*The Co-Stimulatory Signal During T-cell Activation*), CSK (*Activation of Csk by cAMP-Dependent Protein Kinase Inhibits Signaling through the T Cell Receptor*), 41BB (*The 4-1BB-Dependent Immune Response*), NKCELLS (*Ras-Independent Pathway in NK Cell-Mediated Cytotoxicity*), and NO2IL12 (*NO2-Dependent IL12 Pathway in NK Cells*). Overall, these results are consistent with recent reports also stemming from analyses of TCGA transcriptional datasets. Li and colleagues conclude that relative to matching normal tissues, CD8 T cells are highly enriched in the primary tumors of head and neck cancer, and kidney renal clear cell carcinoma (Li et al., 2016). The opposite is true for lung and colorectal cancers. A similar report by Senbabaoglu and colleagues points to kidney renal clear cell carcinoma exhibiting the highest level of a T cell expression signature (Senbabaoglu et al., 2016). Results from our bioinformatic analyses are consistent with those generated via immunohistochemistry, digital imaging, and flow cytometry (Degnim et al., 2017). Through immunohistochemistry, elevated populations of various immune cells such as CD4 T cells, CD8 T cells, B cells (CD20), macrophages (CD68), and dendritic cells (CD11c) have been observed in renal (Michael & Pandha, 2003; Murphy et al., 2015), head and neck (Lei et al., 2016), cervical (Patel & Chiplunkar, 2009), esophageal (Hatogai et al., 2016), prostate (Strasner & Karin, 2015), skin (Rajkumar & Watson, 2016), breast (Luen, Savas, Fox, Salgado, & Loi, 2017; Miyan, Schmidt-Mende, Kiessling, Poschke, & de Boniface, 2016), and thyroid (Cunha et al., 2015) cancers. Within a cancer type, there can be discrepancies in the extent and type of immune infiltration. For example, in breast cancer, immunohistochemical staining indicated that the luminal A subtype (which tends to be ER positive/HER2 negative) appears to be less immunogenic, while the poor prognosis basal subtype (primarily negative in ER, HER2, PR; also known as triple-negative) is commonly populated with FoxP3-expressing T reg cells (Luen et al., 2017).

When clinical outcome is integrated with immunohistochemical results, the positive prognostic values (i.e., higher expression = better prognosis) of T cell surface proteins (such as CD4, CD8) are evident in head neck cancer (Nguyen et al., 2016), melanoma (Erdag et al., 2012; Kakavand et al., 2015), cervical cancer (Nedergaard, Ladekarl, Thomsen, Nyengaard, & Nielsen, 2007; Piersma et al., 2007), bladder cancer (Oguro et al., 2015; Zhang et al., 2017), and liver cancer (Gabrielson et al., 2016). A recent publication by Iglesia and colleagues (using TCGA data) suggests that a more prominent T cell signature is associated with better prognosis in various cancer types

(Iglesia et al., 2016). Our own analyses of TCGA transcriptional datasets also indicate immune-related pathways being highly associated with favorable prognosis in head and neck, bladder, liver, breast, cervical, and ovarian cancers, as well as melanoma, lower grade glioma, and sarcoma. Most notable of these T cell-related pathways are CSK, NO2IL12, IL12, CTLA4, NKCELLS, and TCR. Each of the aforementioned pathways includes several interesting component genes, which by themselves have prognostic values. These include: ZAP70 and LCK (both of which are components of the CSK pathway, or *"Activation of Csk by cAMP-Dependent Protein Kinase Inhibits Signaling Through the T Cell Receptor"*), being positively prognostic (i.e., high expression=better prognosis) for head and neck cancer, melanoma, bladder cancer, and liver cancer. Part of the NO2IL12 pathway (*NO2-Dependent IL12 Pathway in NK Cells*) are the genes IL12B, TYK2, and CD2, whose expression levels are also positively prognostic for head and neck cancer, melanoma, cervical cancer, and breast cancer. The activation of CSK pathway results in T cell proliferation, while that of the NO2IL12 pathway results in Th1 helper T cell differentiation. In the CSK pathway, the intracellular kinase LCK (Lymphocyte-Specific Protein Tyrosine) phosphorylates the T cell receptor complex (TCR), causing the recruitment of ZAP70 (Zeta-Chain-Associated Protein Kinase 70), which then binds to TCR's CD3-zeta chain. ZAP70 will then phosphorylate the LAT protein (Linker of Activated T Cells), and eventual transcriptional activation of genes involved in T cell proliferation (Mustelin & Tasken, 2003). The NO2IL12 pathway starts with a macrophage secreting the cytokine IL12 (IL12B is a subunit), which then stimulates NK cells to activate JAK2 kinase. JAK2 then phosphorylates TYK2 kinase (Non-Receptor Tyrosine-Protein Kinase), which activates the STAT4 transcription factor, followed by IFN gamma transcription (Bogdan, Rollinghoff, & Diefenbach, 2000). IFN gamma will then induce the differentiation of Th1 helper T cells. Previous transcriptional profiling study has shown that ZAP70, LCK, and CD2 are indeed among the most positively prognostic genes in melanoma (Bogunovic et al., 2009). On the other hand, elevated levels of immunohistochemically-detected markers of T cell infiltrates have negative association with prognosis in renal cancers (Giraldo et al., 2015; Mella et al., 2015; Nakano et al., 2001; Thompson et al., 2007). In our analysis of TCGA transcriptional data, we observed that lower expression of markers such ZAP70, CD2, PDCD1, and CD7 is associated with favorable outcome in lower grade gliomas. In renal clear cell carcinoma (KIRC), there is a similar trend for the genes ZAP70, PDCD1, and CD7. The composition of T cell infiltrates is likely a determining factor in how immune infiltrates influence prognosis. As previous studies have shown, T reg cells are associated with an unfavorable outcome in cancer (Tanaka & Sakaguchi, 2017). A marker of T reg cells, the regulatory protein Foxp3, appears to indicate recurrence in lung cancer (Yan, Jiao, Zhang, Guan, & Wang, 2017), and poor prognosis in glioblastoma (Yue et al., 2014). Another gene which according to our analysis is a frequent negative prognostic marker in cancer is BIRC5 (or survivin; an inhibitor of apoptosis). This gene is a prominent component of the *B Cell Survival Pathway* (BCELLSURVIVAL).

As previously demonstrated, the integration of TCGA methylation and expression data can provide clues regarding the epigenetic regulation of any given gene (Bacolod et al., 2015). A similar approach was then applied to some of the important T cell marker genes (such as CTLA4, PD1, CD3 genes). Results suggest that these genes in general are epigenetically regulated, with a cluster of CpG sites influencing their transcription. These bioinformatic observations are consistent with previous reports regarding the epigenetic regulation of PD1 (Goltz et al., 2016; Marwitz et al., 2017), CTLA4 (Marwitz et al., 2017), as well as other genes associated with T cell signaling (Limbach et al., 2016). Moreover, these observations suggest that CpG methylation can possibly serve as surrogate markers for immune-stratification of cancer patients (Bacolod, Barany, & Fisher, 2019).

Although at this point further experimental confirmation, defining pathways and epigenetic regulatory mechanisms predicted to influence immune infiltration in cancer, may eventually enhance immunotherapy of cancer. The reality is that, "although fast becoming a standard in the regimen for clinical management of cancer," the effectiveness of immune checkpoint inhibitors is still limited. For example, in SKCM the response to anti-CTLA-4 immunotherapy (ipilimumab) in patients is only 10.9% (Hodi et al., 2010). Our observations also suggest that certain types of tumors, because of more prominent activation of T cell-related pathways, may be more receptive to immune checkpoint inhibition. This may explain why the current approvals for checkpoint inhibitors are for cancer types considered highly infiltrated by T cells. The anti-PD1 nivolumab and pembrolizumab, and the anti-CTLA4 ipilimumab, are approved by the FDA for treatment of metastatic melanoma (Nakamura & Okuyama, 2016), which as explained above exhibits highly dysregulated pathways associated with infiltration by T lymphocytes, dendritic cells, and NK cells. Also FDA-approved are treatments for metastatic renal cancer (nivolumab) and head and neck cancer (nivolumab and pembrolizumab), two other highly infiltrated tumors.

The other approvals were for lung cancer (nivolumab and pembrolizumab, anti-PDL1 atezolizumab) and urothelial carcinoma (nivolumab, atezolizumab) (Ning et al., 2017; Prasad & Kaestner, 2017). However, with varying degrees of immune infiltration in every cancer type (even those which do not appear to be differentially infiltrated by immune cells, based on bioinformatic analysis), it is expected that there will always be a subset of patients who will respond positively to checkpoint inhibitors. Therefore, there is a need to develop a highly reliable predictive test for these checkpoint inhibitors. As of now, the widely employed predictive test for checkpoint inhibition response is immunohistochemistry for PDL1 (CD274), a surface protein found in an antigen presenting (or cancer) cell (see Gibney, Weiner, & Atkins, 2016 for review). The binding of cancer-cell's PDL1 to activated T cell's PDCD1 is a signal which inhibits T cell proliferation. However, results from a clinical trial involving 452 melanoma patients indicated that the association of pre-treated PDL1 expression (measured using immunohistochemistry) with response to pembrolizumab is far from desirable (Daud et al., 2016). According to their report, even PDL1-negative cases may achieve positive response to the drug. Other biomarkers which may predict improved clinical outcome (as reported in various studies) are: mutation rate of the tumor (which can be measured by exome sequencing); elevated density of CD8+ tumor-infiltrating lymphocytes (through immunohistochemistry); tumor microenvironment immune gene signatures (through expression profiling). Ock and colleagues have proposed a simple stratification of tumors based on PDL1 and CD8 expression, with Type I (PDL1 high, CD8 high) found to be most frequent in lung adenocarcinoma and renal clear cell carcinoma (Ock et al., 2016). The Type I group also happens to have more frequent neoantigen mutation rates in various cancer types including lung, bladder, breast, melanoma (Chen et al., 2017). In addition, high PDL1 expression may be driven by increased copy number of the gene (Budczies et al., 2017).

The predicted epigenetic regulation of certain genes crucial to T cell proliferation (such as CD3 genes, PDCD1, CTLA4) may be relevant to the search for patient immune-stratification markers. The concerned CpG sites can serve as surrogate markers in a manner similar to the way MGMT and MLH1 markers are assayed in clinics through quantitation of specific CpG sites at their promoter regions (Bettstetter et al., 2007; Wick et al., 2014). The epigenetic regulation of the aforementioned T cell genes may also be a factor to consider given the profound interest in treating cancer by combining immune checkpoint inhibition with epigenetic drugs such as the hypomethylating agent (or DNMT inhibitor) decitabine, and histone deacetylase inhibitors entinostat and vorinostat (Weintraub, 2016). Among the ongoing clinical trials are: nivolumab/decitabine combination to treat non-small cell lung cancer (NSCLC) (NCT02664181); pembrolizumab/decitabine combination against HER2-negative breast cancer (NCT02957968), and NSCLC (NCT03233724); pembrolizumab/entinostat combination against relapsed and refractory lymphomas (NCT03179930); pembrolizumab/vorinostat combination against Stage IV NSCLC (NCT02638090).

In summary, we demonstrate that certain types of solid tumors appear to exhibit higher tendencies for activated immune-related pathways (particularly those related to infiltration and activation of cytotoxic and helper T cells, dendritic cells, and natural killer cells), and that the activation of certain genes crucial to these pathways may be epigenetically regulated. A deeper understanding of the differences (and similarities) between cancer types (as far as immune infiltration is concerned) may enhance the quest to improve immunotherapy, via effective patient immune-stratification, or new immune-therapeutic targets.

5. Expert opinion

The similarities and differences across cancer types can now be appreciated in molecular terms (e.g., molecular biomarkers, dysregulated pathways, epigenetics). This report, which employed integrated analysis of publicly available cancer genomic datasets (primary tumors vs. normal cells, except melanoma wherein metastatic samples were compared to primary tumors), further validates this widely accepted principle. However, a closer examination of the results indicates that these molecular differences and similarities are not just confined to cancer cells themselves, but also apply to infiltrating immune cells. For certain cancer types, such as melanoma, renal cancers, and head and neck cancer, there seems to be a higher degree of relative immune infiltration (and relative activation of T cells and other immune signaling pathways). Perhaps, these observations reinforce the expected limitations of checkpoint inhibition immunotherapy (especially when used in treating primary cases): that effectiveness may be constrained to cancer cases with relatively high immune infiltration that are less prevalent in certain cancer types. However, deeper understanding of the processes governing the activation of cancer-killing immune cells may also allow us to develop new or complementary drugs which can circumvent these limitations. Achieving these objectives may result in a further refinement in the immunotherapy of cancer and an ability to personalize medical treatments to enhance therapeutic outcomes.

Acknowledgments
We would like to thank: Earlier.org and Acuamark Diagnostics for their financial support. P.B.F. acknowledges the continued generous support from the National Foundation for Cancer Research (NFCR).

References
Bacolod, M. D., Barany, F., & Fisher, P. B. (2019). Can CpG methylation serve as potential surrogate markers for immune infiltration in cancer? *Advances in Cancer Research, 143*. [in press].

Bacolod, M. D., Das, S. K., Sokhi, U. K., Bradley, S., Fenstermacher, D. A., & Pellecchia, M. (2015). Examination of epigenetic and other molecular factors associated with mda-9/Syntenin dysregulation in cancer through integrated analyses of public genomic datasets. *Advances in Cancer Research, 127*, 49–121.

Bacolod, M. D., Talukdar, S., Emdad, L., Das, S. K., Sarkar, D., & Wang, X.-Y. (2016). Immune infiltration, glioma stratification, and therapeutic implications. *Translational Cancer Research, 5*, S652–S656.

Barrett, T., & Edgar, R. (2006). Gene expression omnibus: Microarray data storage, submission, retrieval, and analysis. *Methods in Enzymology, 411*, 352–369.

Berger, M. F., & Mardis, E. R. (2018). The emerging clinical relevance of genomics in cancer medicine. *Nature Reviews. Clinical Oncology, 15*(6), 353–365.

Bernstein, B. E., Kamal, M., Lindblad-Toh, K., Bekiranov, S., Bailey, D. K., & Huebert, D. J. (2005). Genomic maps and comparative analysis of histone modifications in human and mouse. *Cell, 120*(2), 169–181.

Bethmann, D., Feng, Z., & Fox, B. A. (2017). Immunoprofiling as a predictor of patient's response to cancer therapy—Promises and challenges. *Current Opinion in Immunology, 45*, 60–72.

Bettstetter, M., Dechant, S., Ruemmele, P., Grabowski, M., Keller, G., & Holinski-Feder,-E. (2007). Distinction of hereditary nonpolyposis colorectal cancer and sporadic microsatellite-unstable colorectal cancer through quantification of MLH1 methylation by real-time PCR. *Clinical Cancer Research, 13*(11), 3221–3228.

Bogdan, C., Rollinghoff, M., & Diefenbach, A. (2000). The role of nitric oxide in innate immunity. *Immunological Reviews, 173*, 17–26.

Bogunovic, D., O'Neill, D. W., Belitskaya-Levy, I., Vacic, V., Yu, Y. L., & Adams, S. (2009). Immune profile and mitotic index of metastatic melanoma lesions enhance clinical staging in predicting patient survival. *Proceedings of the National Academy of Sciences of the United States of America, 106*(48), 20429–20434.

Bondanza, A., & Casucci, M. (2016). Tumor immunology methods and protocols. In *Methods in molecular biology*. New York; Heidelberg: Humana Press.

Boyle, A. P., Davis, S., Shulha, H. P., Meltzer, P., Margulies, E. H., & Weng, Z. (2008). High-resolution mapping and characterization of open chromatin across the genome. *Cell, 132*(2), 311–322.

Budczies, J., Bockmayr, M., Klauschen, F., Endris, V., Frohling, S., & Schirmacher, P. (2017). Mutation patterns in genes encoding interferon signaling and antigen presentation: A pan-cancer survey with implications for the use of immune checkpoint inhibitors. *Genes, Chromosomes and Cancer, 56*(8), 651–659.

Cai, B., & Jiang, X. (2014). Revealing biological pathways implicated in lung cancer from TCGA gene expression data using gene set enrichment analysis. *Cancer Informatics, 13*(Suppl. 1), 113–121.

Campbell, M. J., & Turner, B. M. (2013). Altered histone modifications in cancer. *Advances in Experimental Medicine and Biology, 754*, 81–107.

Chen, M., Rothman, N., Ye, Y., Gu, J., Scheet, P. A., & Huang, M. (2016). Pathway analysis of bladder cancer genome-wide association study identifies novel pathways involved in bladder cancer development. *Genes & Cancer, 7*(7–8), 229–239.

Chen, Y. P., Zhang, Y., Lv, J. W., Li, Y. Q., Wang, Y. Q., & He, Q. M. (2017). Genomic analysis of tumor microenvironment immune types across 14 solid cancer types: Immunotherapeutic implications. *Theranostics, 7*(14), 3585–3594.

Couzin-Frankel, J. (2013). Breakthrough of the year 2013. Cancer immunotherapy. *Science, 342*(6165), 1432–1433.

Cox, D. R. (1972). Regression models and life-tables. *Journal of the Royal Statistical Society. Series B, Statistical Methodology, 34*(2), 187–220.

Cunha, L. L., Marcello, M. A., Nonogaki, S., Morari, E. C., Soares, F. A., & Vassallo, J. (2015). CD8+ tumour-infiltrating lymphocytes and COX2 expression may predict relapse in differentiated thyroid cancer. *Clinical Endocrinology, 83*(2), 246–253.

Danaher, P., Warren, S., Dennis, L., D'Amico, L., White, A., & Disis, M. L. (2017). Gene expression markers of tumor infiltrating leukocytes. *Journal for Immunotherapy of Cancer, 5*(18), 1–15.

Daud, A. I., Wolchok, J. D., Robert, C., Hwu, W. J., Weber, J. S., & Ribas, A. (2016). Programmed death-ligand 1 expression and response to the anti-programmed death 1 antibody pembrolizumab in melanoma. *Journal of Clinical Oncology, 34*(34), 4102–4109.

Degnim, A. C., Hoskin, T. L., Arshad, M., Frost, M. H., Winham, S. J., & Brahmbhatt, R. A. (2017). Alterations in the immune cell composition in premalignant breast tissue that precede breast cancer development. *Clinical Cancer Research, 23*(14), 3945–3952.

D'Errico, G., Machado, H. L., & Sainz, B., Jr. (2017). A current perspective on cancer immune therapy: Step-by-step approach to constructing the magic bullet. *Clinical and Translational Medicine, 6*, 3.

Efron, B., & Tibshirani, R. (2007). On testing the significance of sets of genes. *Annals of Applied Statistics, 1*(1), 107–129.

ENCODE Project Consortium. (2004). The ENCODE (ENCyclopedia Of DNA Elements) project. *Science, 306*(5696), 636–640.

Erdag, G., Schaefer, J. T., Smolkin, M. E., Deacon, D. H., Shea, S. M., & Dengel, L. T. (2012). Immunotype and immunohistologic characteristics of tumor-infiltrating immune cells are associated with clinical outcome in metastatic melanoma. *Cancer Research, 72*(5), 1070–1080.

Ernst, J., Kheradpour, P., Mikkelsen, T. S., Shoresh, N., Ward, L. D., & Epstein, C. B. (2011). Mapping and analysis of chromatin state dynamics in nine human cell types. *Nature, 473*(7345), 43–49.

Fridman, W. H., Pages, F., Sautes-Fridman, C., & Galon, J. (2012). The immune contexture in human tumours: Impact on clinical outcome. *Nature Reviews Cancer, 12*(4), 298–306.

Gabrielson, A., Wu, Y., Wang, H., Jiang, J., Kallakury, B., & Gatalica, Z. (2016). Intratumoral CD3 and CD8 T-cell densities associated with relapse-free survival in HCC. *Cancer Immunology Research, 4*(5), 419–430.

Ghirlando, R., Giles, K., Gowher, H., Xiao, T., Xu, Z., & Yao, H. (2012). Chromatin domains, insulators, and the regulation of gene expression. *Biochimica et Biophysica Acta, 1819*(7), 644–651.

Gibney, G. T., Weiner, L. M., & Atkins, M. B. (2016). Predictive biomarkers for checkpoint inhibitor-based immunotherapy. *Lancet Oncology, 17*(12), e542–e551.

Giraldo, N. A., Becht, E., Pages, F., Skliris, G., Verkarre, V., & Vano, Y. (2015). Orchestration and prognostic significance of immune checkpoints in the microenvironment of primary and metastatic renal cell Cancer. *Clinical Cancer Research, 21*(13), 3031–3040.

Giraldo, N. A., Becht, E., Remark, R., Damotte, D., Sautes-Fridman, C., & Fridman, W. H. (2014). The immune contexture of primary and metastatic human tumours. *Current Opinion in Immunology, 27*, 8–15.

Glanzer, J. G., & Eberwine, J. H. (2004). Expression profiling of small cellular samples in cancer: Less is more. *British Journal of Cancer, 90*(6), 1111–1114.

Goldman, M., Craft, B., Swatloski, T., Ellrott, K., Cline, M., & Diekhans, M. (2013). The UCSC Cancer Genomics Browser: Update 2013. *Nucleic Acids Research, 41*, D949–D954. [Database issue].

Goltz, D., Gevensleben, H., Dietrich, J., Ellinger, J., Landsberg, J., & Kristiansen, G. (2016). Promoter methylation of the immune checkpoint receptor PD-1 (PDCD1) is an independent prognostic biomarker for biochemical recurrence-free survival in prostate cancer patients following radical prostatectomy. *Oncoimmunology, 5*(10), e1221555.

Greil, R., Hutterer, E., Hartmann, T. N., & Pleyer, L. (2017). Reactivation of dormant antitumor immunity—A clinical perspective of therapeutic immune checkpoint modulation. *Cell Communication and Signaling, 15*, 5.

Hatogai, K., Kitano, S., Fujii, S., Kojima, T., Daiko, H., & Nomura, S. (2016). Comprehensive immunohistochemical analysis of tumor microenvironment immune status in esophageal squamous cell carcinoma. *Oncotarget, 7*(30), 47252–47264.

Hodi, F. S., O'Day, S. J., McDermott, D. F., Weber, R. W., Sosman, J. A., & Haanen, J. B. (2010). Improved survival with ipilimumab in patients with metastatic melanoma. *New England Journal of Medicine, 363*(8), 711–723.

Iglesia, M. D., Parker, J. S., Hoadley, K. A., Serody, J. S., Perou, C. M., & Vincent, B. G. (2016). Genomic analysis of immune cell infiltrates across 11 tumor types. *Journal of the National Cancer Institute, 108*, 11.

Iorio, F., Knijnenburg, T. A., Vis, D. J., Bignell, G. R., Menden, M. P., Schubert, M., et al. (2016). A landscape of pharmacogenomic interactions in cancer. *Cell, 166*(3), 740–754.

Kaiser, J. (2005). National Institutes of Health. NCI gears up for cancer genome project. *Science, 307*(5713), 1182.

Kakavand, H., Vilain, R. E., Wilmott, J. S., Burke, H., Yearley, J. H., & Thompson, J. F. (2015). Tumor PD-L1 expression, immune cell correlates and PD-1+ lymphocytes in sentinel lymph node melanoma metastases. *Modern Pathology, 28*(12), 1535–1544.

Lei, Y., Xie, Y., Tan, Y. S., Prince, M. E., Moyer, J. S., & Nor, J. (2016). Telltale tumor infiltrating lymphocytes (TIL) in oral, head & neck cancer. *Oral Oncology, 61*, 159–165.

Li, B., Severson, E., Pignon, J. C., Zhao, H., Li, T., & Novak, J. (2016). Comprehensive analyses of tumor immunity: Implications for cancer immunotherapy. *Genome Biology, 17*, 174.

Limbach, M., Saare, M., Tserel, L., Kisand, K., Eglit, T., & Sauer, S. (2016). Epigenetic profiling in CD4+ and CD8+ T cells from Graves' disease patients reveals changes in genes associated with T cell receptor signaling. *Journal of Autoimmunity, 67*, 46–56.

Luen, S. J., Savas, P., Fox, S. B., Salgado, R., & Loi, S. (2017). Tumour-infiltrating lymphocytes and the emerging role of immunotherapy in breast cancer. *Pathology, 49*(2), 141–155.

Marwitz, S., Scheufele, S., Perner, S., Reck, M., Ammerpohl, O., & Goldmann, T. (2017). Epigenetic modifications of the immune-checkpoint genes CTLA4 and PDCD1 in nonsmall cell lung cancer results in increased expression. *Clinical Epigenetics, 9*, 51.

Mella, M., Kauppila, J. H., Karihtala, P., Lehenkari, P., Jukkola-Vuorinen, A., & Soini, Y. (2015). Tumor infiltrating CD8(+) T lymphocyte count is independent of tumor TLR9 status in treatment naive triple negative breast cancer and renal cell carcinoma. *Oncoimmunology, 4*(6), e1002726.

Michael, A., & Pandha, H. S. (2003). Renal-cell carcinoma: Tumour markers, T-cell epitopes, and potential for new therapies. *Lancet Oncology, 4*(4), 215–223.

Miyan, M., Schmidt-Mende, J., Kiessling, R., Poschke, I., & de Boniface, J. (2016). Differential tumor infiltration by T-cells characterizes intrinsic molecular subtypes in breast cancer. *Journal of Translational Medicine, 14*, 227.

Murphy, K. A., James, B. R., Guan, Y., Torry, D. S., Wilber, A., & Griffith, T. S. (2015). Exploiting natural anti-tumor immunity for metastatic renal cell carcinoma. *Human Vaccines & Immunotherapeutics, 11*(7), 1612–1620.

Mustelin, T., & Tasken, K. (2003). Positive and negative regulation of T-cell activation through kinases and phosphatases. *Biochemical Journal, 371*(Pt. 1), 15–27.

Nakamura, K., & Okuyama, R. (2016). Immunotherapy for advanced melanoma: Current knowledge and future directions. *Journal of Dermatological Science, 83*(2), 87–94.

Nakano, O., Sato, M., Naito, Y., Suzuki, K., Orikasa, S., & Aizawa, M. (2001). Proliferative activity of intratumoral CD8(+) T-lymphocytes as a prognostic factor in human renal cell carcinoma: Clinicopathologic demonstration of antitumor immunity. *Cancer Research, 61*(13), 5132–5136.

Narrandes, S., Huang, S., Murphy, L., & Xu, W. (2018). The exploration of contrasting pathways in triple negative breast cancer (TNBC). *BMC Cancer, 18*, 22.

Nedergaard, B. S., Ladekarl, M., Thomsen, H. F., Nyengaard, J. R., & Nielsen, K. (2007). Low density of CD3+, CD4+ and CD8+ cells is associated with increased risk of relapse in squamous cell cervical cancer. *British Journal of Cancer, 97*(8), 1135–1138.

Nguyen, N., Bellile, E., Thomas, D., McHugh, J., Rozek, L., & Virani, S. (2016). Tumor infiltrating lymphocytes and survival in patients with head and neck squamous cell carcinoma. *Head and Neck, 38*(7), 1074–1084.

Nguyen, D., & Gocke, C. D. (2017). Managing the genomic revolution in cancer diagnostics. *Virchows Archiv, 471*(2), 175–194.

Ning, B., Su, Z., Mei, N., Hong, H., Deng, H., & Shi, L. (2014). Toxicogenomics and cancer susceptibility: Advances with next-generation sequencing. *Journal of Environmental Science and Health, Part C: Environmental Carcinogenesis and Ecotoxicology Reviews, 32*(2), 121–158.

Ning, Y. M., Suzman, D., Maher, V. E., Zhang, L., Tang, S., & Ricks, T. (2017). FDA approval summary: Atezolizumab for the treatment of patients with progressive advanced urothelial carcinoma after platinum-containing chemotherapy. *The Oncologist, 22*(6), 743–749.

Ock, C. Y., Keam, B., Kim, S., Lee, J. S., Kim, M., & Kim, T. M. (2016). Pan-cancer immunogenomic perspective on the tumor microenvironment based on PD-L1 and CD8 T-cell infiltration. *Clinical Cancer Research, 22*(9), 2261–2270.

Oguro, S., Ino, Y., Shimada, K., Hatanaka, Y., Matsuno, Y., & Esaki, M. (2015). Clinical significance of tumor-infiltrating immune cells focusing on BTLA and Cbl-b in patients with gallbladder cancer. *Cancer Science, 106*(12), 1750–1760.

Patel, S., & Chiplunkar, S. (2009). Host immune responses to cervical cancer. *Current Opinion in Obstetrics and Gynecology, 21*(1), 54–59.

Piersma, S. J., Jordanova, E. S., van Poelgeest, M. I., Kwappenberg, K. M., van der Hulst, J. M., & Drijfhout, J. W. (2007). High number of intraepithelial CD8+ tumor-infiltrating lymphocytes is associated with the absence of lymph node metastases in patients with large early-stage cervical cancer. *Cancer Research, 67*(1), 354–361.

Player, A., Barrett, J. C., & Kawasaki, E. S. (2004). Laser capture microdissection, microarrays and the precise definition of a cancer cell. *Expert Review of Molecular Diagnostics, 4*(6), 831–840.

Prasad, V., & Kaestner, V. (2017). Nivolumab and pembrolizumab: Monoclonal antibodies against programmed cell death-1 (PD-1) that are interchangeable. *Seminars in Oncology, 44*(2), 132–135.

Rajkumar, S., & Watson, I. R. (2016). Molecular characterisation of cutaneous melanoma: Creating a framework for targeted and immune therapies. *British Journal of Cancer, 115*(2), 145–155.

Rooney, M. S., Shukla, S. A., Wu, C. J., Getz, G., & Hacohen, N. (2015). Molecular and genetic properties of tumors associated with local immune cytolytic activity. *Cell*, *160*(1–2), 48–61.

Sandoval, J., Heyn, H., Moran, S., Serra-Musach, J., Pujana, M. A., & Bibikova, M. (2011). Validation of a DNA methylation microarray for 450, 000 CpG sites in the human genome. *Epigenetics*, *6*(6), 692–702.

Senbabaoglu, Y., Gejman, R. S., Winer, A. G., Liu, M., Van Allen, E. M., & de Velasco, G. (2016). Tumor immune microenvironment characterization in clear cell renal cell carcinoma identifies prognostic and immunotherapeutically relevant messenger RNA signatures. *Genome Biology*, *17*, 231.

Senft, D., Leiserson, M. D. M., Ruppin, E., & Ronai, Z. A. (2017). Precision oncology: The road ahead. *Trends in Molecular Medicine*, *23*(10), 874–898.

Siemers, N. O., Holloway, J. L., Chang, H., Chasalow, S. D., Ross-Mac Donald, P. B., & Voliva, C. F. (2017). Genome-wide association analysis identifies genetic correlates of immune infiltrates in solid tumors. *PLoS One*, *12*(7), e0179726.

Simon, R., Lam, A., Li, M. C., Ngan, M., Menenzes, S., & Zhao, Y. (2007). Analysis of gene expression data using BRB-Array Tools. *Cancer Informatics*, *3*, 11–17.

Stadler, Z. K., Schrader, K. A., Vijai, J., Robson, M. E., & Offit, K. (2014). Cancer genomics and inherited risk. *Journal of Clinical Oncology*, *32*(7), 687–698.

Strasner, A., & Karin, M. (2015). Immune infiltration and prostate cancer. *Frontiers in Oncology*, *5*, 128.

Subramanian, A., Tamayo, P., Mootha, V. K., Mukherjee, S., Ebert, B. L., & Gillette, M. A. (2005). Gene set enrichment analysis: A knowledge-based approach for interpreting genome-wide expression profiles. *Proceedings of the National Academy of Sciences of the United States of America*, *102*(43), 15545–15550.

Sukari, A., Nagasaka, M., Al-Hadidi, A., & Lum, L. G. (2016). Cancer immunology and immunotherapy. *Anticancer Research*, *36*(11), 5593–5606.

Sweis, R. F., & Luke, J. J. (2017). Mechanistic and pharmacologic insights on immune checkpoint inhibitors. *Pharmacological Research*, *120*, 1–9.

Tanaka, A., & Sakaguchi, S. (2017). Regulatory T cells in cancer immunotherapy. *Cell Research*, *27*(1), 109–118.

Thompson, R. H., Dong, H., Lohse, C. M., Leibovich, B. C., Blute, M. L., & Cheville, J. C. (2007). PD-1 is expressed by tumor-infiltrating immune cells and is associated with poor outcome for patients with renal cell carcinoma. *Clinical Cancer Research*, *13*(6), 1757–1761.

Ursini-Siegel, J., & Beauchemin, N. (2016). The tumor microenvironment methods and protocols. In *Methods in molecular biology*. New York: Humana Press.

Weintraub, K. (2016). Take two: Combining immunotherapy with epigenetic drugs to tackle cancer. *Nature Medicine*, *22*(1), 8–10.

Wick, W., Weller, M., van den Bent, M., Sanson, M., Weiler, M., & von Deimling, A. (2014). MGMT testing—The challenges for biomarker-based glioma treatment. *Nature Reviews Neurology*, *10*(7), 372–385.

Yan, X., Jiao, S. C., Zhang, G. Q., Guan, Y., & Wang, J. L. (2017). Tumor-associated immune factors are associated with recurrence and metastasis in non-small cell lung cancer. *Cancer Gene Therapy*, *24*(2), 57–63.

Yue, Q., Zhang, X., Ye, H. X., Wang, Y., Du, Z. G., & Yao, Y. (2014). The prognostic value of Foxp 3 + tumor-infiltrating lymphocytes in patients with glioblastoma. *Journal of Neurooncology*, *116*(2), 251–259.

Zhang, Y., Ma, C., Wang, M., Hou, H., Cui, L., & Jiang, C. (2017). Prognostic significance of immune cells in the tumor microenvironment and peripheral blood of gallbladder carcinoma patients. *Clinical and Translational Oncology*, *19*(4), 477–488.

Zhu, J., Sanborn, J. Z., Benz, S., Szeto, C., Hsu, F., & Kuhn, R. M. (2009). The UCSC Cancer Genomics Browser. *Nature Methods*, *6*(4), 239–240.

CHAPTER FIVE

HVEM network signaling in cancer

John R. Šedý*, Parham Ramezani-Rad

Infectious and Inflammatory Disease Center, NCI-Designated Cancer Center, Sanford Burnham Prebys Medical Discovery Institute, La Jolla, CA, United States
*Corresponding author: e-mail address: jsedy@sbpdiscovery.org

Contents

1. Introduction	148
2. HVEM network interactions in the immune microenvironment	149
3. Genetic lesions in *TNFRSF14* in lymphoid cancers	151
3.1 Immune microenvironment within lymphoma and influence of HVEM ligands	157
4. Genetic lesions in *TNFRSF14* in non-lymphoid cancers	158
5. HVEM functions within the tumor microenvironment	160
5.1 LTα and LIGHT activate pleiotropic anti-tumor functions	161
5.2 BTLA identifies exhausted T cells	163
5.3 CD160 co-stimulates anti-tumor IFNγ	165
5.4 Intracellular signaling pathways activated by HVEM	166
6. Therapeutic targeting of the HVEM network in lymphoma and other tumors	169
6.1 LTα	169
6.2 LIGHT	170
6.3 BTLA	171
6.4 CD160	172
7. Concluding remarks	172
Funding	173
References	173

Abstract

Somatic mutations in cancer cells may influence tumor growth, survival, or immune interactions in their microenvironment. The tumor necrosis factor receptor family member HVEM (*TNFRSF14*) is frequently mutated in cancers and has been attributed a tumor suppressive role in some cancer contexts. HVEM functions both as a ligand for the lymphocyte checkpoint proteins BTLA and CD160, and as a receptor that activates NF-κB signaling pathways in response to BTLA and CD160 and the TNF ligands LIGHT and LTα. BTLA functions to inhibit lymphocyte activation, but has also been ascribed a role in stimulating cell survival. CD160 functions to co-stimulate lymphocyte function, but has also been shown to activate inhibitory signaling in CD4$^+$ T cells. Thus, the role of HVEM within diverse cancers and in regulating the immune responses to these tumors is likely context specific. Additionally, development of therapeutics that target proteins within this network of interacting proteins will require a deeper understanding of how these proteins function in a cancer-specific manner. However, the prominent role of the HVEM network in anti-cancer immune responses indicates a promising area for drug development.

Abbreviations

2B4	CD244
ABC	activated B cell
AKT	AKT serine/threonine kinase 1, AKT1
AML	acute myeloid leukemia
ARID1A	AT-rich interaction domain 1A
AXL	Axl receptor tyrosine kinase
B2M	beta-2-microglobulin
B-ALL	B cell acute lymphocytic leukemia
B-CLL	B cell chronic lymphocytic leukemia
BCL	B cell lymphoma (2; 6)
BCR	B cell receptor
BIRC	baculoviral inhibitor-of-apoptosis repeat containing (2; 3)
BL	Burkitt lymphoma
BRCA	breast cancer (TCGA)
Breg	regulatory B cell
BTK	Bruton's tyrosine kinase
BTLA	B and T-lymphocyte associated, CD272
CARD11	caspase-associated recruitment domain 11
CAR-T cell	chimeric antigen receptor-T cell
CEBPA	CCAAT enhancer binding protein alpha
CIITA	class II major histocompatibility complex transactivator
CLL	chronic lymphocytic leukemia
CMV	cytomegalovirus
CNS	central nervous system
COCA	colon cancer (TCGA)
COSMIC	catalog of somatic mutations in cancer
CRD	cysteine-rich domain
CREBBP	c-AMP response element binding protein binding protein
CRISPR	clustered regularly interspaced short palindromic repeats
CSCC	cutaneous squamous cell carcinoma (TCGA)
CTCL	cutaneous T cell lymphoma
CTLA-4	cytotoxic T-lymphocyte associated protein 4, CD152
DC	dendritic cell
DcR3	decoy receptor 3, TNFRSF6B
DIM	detergent-insoluble lipid microdomain
DLBCL	diffuse large B cell lymphoma
EBV	Epstein-Barr virus
EP300	adenovirus E1A-associated cellular p300 transcriptional co-activator protein
ERK	extracellular signal-regulated kinase (1; 2)
EZH2	enhancer of zeste 2 polycomb repressive complex 2 subunit

FAS	Fas cell surface death receptor, TNFRSF6, CD95
FL	follicular lymphoma
FLT3-ITD	Fms related tyrosine kinase 3-internal tandem duplication
FOXO1	forkhead box O1
FOXP3	forkhead box P3
GβL	mTOR-associated protein, LST8 homolog, MLST8
GC	germinal center (B, B cell)
GM-CSF	colony stimulating factor 2, CSF2
GRB2	growth receptor bound 2
GVHD	graft *versus* host disease
GZMB	granzyme B
HCC	hepatocellular carcinoma (TCGA)
HGBL	high-grade B cell lymphoma
HLA-G	major histocompatibility complex, class I, G
HSC	hematopoietic stem cells
HSV	herpes simplex virus
HVEM	herpesvirus entry mediator, TNFRSF14, CD270
ICAM1	intracellular adhesion molecule 1, CD54
ICGC	international cancer genome consortium
IDO	indoleamine 2,3-dioxygenase 1, IDO1
IFNγ	interferon γ, IFNG
IL	interleukin (1β, 2, 4, 6, 7, 10, 12, 18, 21)
ITK	IL2-inducible T cell kinase
ITPKB	inositol-trisphosphate 3-kinase B
JAK	Janus kinase
JNK	mitogen-activated protein kinase 8 (MAPK8)
KMT2D	lysine methyltransferase 2D
LAG-3	lymphocyte-activation gene 3
LIGHT	homologous to lymphotoxin, inducible expression, competes with HSV glycoprotein D for binding to HVEM, expressed on T cells, TNFSF14, CD258
LTα	lymphotoxin (α, LTA, TNFSF1; β, LTB, TNFSF3)
LTβR	lymphotoxin-β receptor, LTBR, TNFRSF3
LUAD	lung adenocarcinoma (TCGA)
MALT	mucosa-associated lymphoid tissue
MALY	malignant lymphoma (TCGA)
MAP2K1	mitogen-activated protein kinase kinase 1
MAPK	mitogen-activated protein kinase
MDSC	myeloid derived suppressor cell
MEF2B	myocyte enhancer factor 2B
MELA	melanoma (TCGA)
MHC	major histocompatibility complex
miR-28	microRNA 28
mTOR	mechanistic target of rapamycin kinase
mTORC2	mechanistic target of rapamycin complex 2
MYD88	myeloid differentiation primary response gene 88
NSCLC	non-small cell lung cancer (TCGA)

NF-κB	nuclear factor kappa-light-chain-enhancer of activated B cells
NIK	NF-κB inducing kinase, MAP3K14
NK	natural killer cell
NKT	natural killer T cell
NMZL	nodal marginal zone lymphoma
NPM1	nucleophosmin 1
OTUD7B	OTU deubiquitinase 7B
p38	mitogen-activated protein kinase 14, MAPK14
PD-1	programmed cell death-1, PDCD1 CD279
PD-L1	programmed cell death 1 ligand 1, CD274
PI3K	phosphoinositide-3-kinase
PRF1	perforin 1
PTEN	phosphatase and tensin homolog
PU.1	Spi-1 proto-oncogene, SPI1
REG3γ	regenerating family member 3γ, REG3G
RELB	RELB proto-oncogene, NF-κB subunit
ROCK1	rho-associated coiled-coil containing protein kinase 1
SALM5	synaptic adhesion-like molecule 5, LRFN5
SHIP	Src homology domain-containing inositol-5-phosphatase, INPP5D
SHP	Src homology domain-containing phosphatase (1, PTPN6; 2, PTPN11)
SIN1	mitogen-activated protein kinase-associated protein 1, MAPKAP1
SMAC	second mitochondrial-derived activator of caspases
STAT	signal transducer and activator of transcription (3; 6)
TCGA	the cancer genome atlas
TCR	T cell receptor
Tfh	follicular T cell
tFL	transformed follicular lymphoma
TGFβ	transforming growth factor β1, TGFB1
TIL	tumor infiltrating lymphocytes
TIM-3	T cell immunoglobulin mucin 3, HAVCR2
TLS	tertiary lymphoid structure
TNF	tumor necrosis factor, TNFSF2
TNFAIP3	TNF alpha-induced protein 3
TNFR	tumor necrosis factor receptor
TRAF	TNF receptor-associated factor (2; 3)
Treg	regulatory T cell
TSLP	thymic stromal lymphopoietin
ZBTB16	zinc finger and BTB containing protein 16

1. Introduction

Significant advances have been made in understanding and treatment of cancer in the last couple of decades through the characterization of anti-cancer immune responses and the advent of checkpoint blockade therapy. Blockade of CTLA-4 and PD-1 inhibition potently activates immune responses for many cancers resulting in remission. However, many patients and non-inflamed (cold) tumors are yet unresponsive to these therapies, necessitating the identification of additional lymphocyte targets that trigger the activation of tumor-reactive immune cells, or of novel oncogenic pathways in tumors that can be targeted directly with oncotherapy. Recent genomic studies have confirmed recurrent genetic lesions associated with particular cancers and have revealed cellular pathways that are frequently dysregulated in tumors. This data have been used to develop prognostic factors to assess disease outcome.

The TNFR superfamily member HVEM (*TNFRSF14*) is a focal point for manipulation by viral pathogens, and mutation in cancers and autoimmunity (Alcami, 2003; Basso & Dalla-Favera, 2015; Callahan, Wolchok, Allison, & Sharma, 2013; Murphy & Murphy, 2010; Pardoll, 2012; Sedy, Bekiaris, & Ware, 2014). A number of gene and expression polymorphisms have been recently identified within the HVEM network of interacting cell surface receptors ligands, highlighting the importance of this network to immunity. These genetic alterations may regulate HVEM network interactions and signaling, potentially providing a selective advantage to cancer. Notably, *TNFRSF14* is frequently deleted or mutated in human lymphoma, with some studies indicating that HVEM may suppress lymphoma development (Bjordahl, Steidl, Gascoyne, & Ware, 2013; Boice et al., 2016). Here, we will review recent advances that implicate the HVEM network in cancer and in anti-tumor immune responses. The molecular and cellular interactions between HVEM and its partner receptors require context-specific analysis to determine how dysregulation of this network may function in cancer and immunity.

2. HVEM network interactions in the immune microenvironment

HVEM interacts with multiple ligands expressed in the immune system including the TNF superfamily cytokines, LTα and LIGHT and the immunoglobulin superfamily members BTLA and CD160 that serve as checkpoint regulators (Fig. 1) (Ware & Sedy, 2011). Recently, the neuronal cell surface receptor SALM5 was identified as a fifth ligand that functions with myeloid-expressed HVEM to regulate neuroinflammation in a mouse model of multiple sclerosis (Zhu et al., 2016). LIGHT binds HVEM at its CRD2 and 3 regions, characteristic of TNF ligand interactions with TNFR proteins (Bodmer, Schneider, & Tschopp, 2002; Murphy & Murphy, 2010).

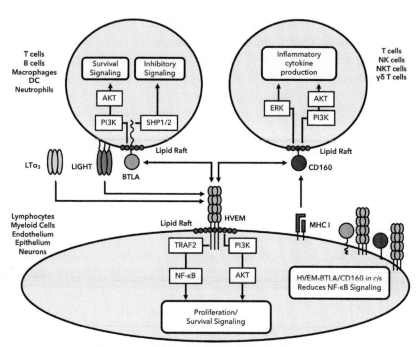

Fig. 1 Major signaling pathways activated by HVEM network proteins. Intermolecular interactions between HVEM and its ligands are indicated between cells. BTLA (upper left) is activated by cell-expressed HVEM, resulting in phosphorylation of its immunoreceptor tyrosine-based inhibitory motif and recruitment of SHP1/2 proteins that dephosphorylate target proteins. BTLA localization to DIMs may facilitate association with PI3K proteins and activation of AKT survival signaling in T cells. CD160 (upper right) is activated by cell-expressed HVEM or MHC I proteins. The predominant form of CD160 is tethered to the cell membrane through a glycosylphosphatidylinositol anchor. CD160 ligation activates PI3K/AKT and ERK1/2 signaling pathways to co-stimulate cytokine production. Human HVEM (bottom) is activated by LIGHT, LTα, BTLA, and CD160. HVEM ligation results in receptor oligomerization and recruitment of TRAF2 to the cytoplasmic TRAF domain of HVEM and downstream NF-κB signaling. HVEM localization to DIMs may promote its association with PI3K proteins and activation of AKT signaling. HVEM complexed with BTLA or CD160 *in cis* on the cell surface prevents efficient ligand-activated signals.

Human LTα also has weak affinity for HVEM while mouse LTα does not bind HVEM, indicating that caution should be taken in translating animal studies to human (Bossen et al., 2006). BTLA and CD160 both bind HVEM at its CRD1 domain, at a topologically distinct surface from the TNF ligands (Kojima, Kajikawa, Shiroishi, Kuroki, & Maenaka, 2011). The leucine-rich repeat domain of SALM5 was also required to bind the CRD1 of HVEM. HVEM is widely expressed in hematopoietic lineage cells, endothelial and epithelial cells, and in neurons, while LIGHT, BTLA and CD160 show more restricted expression to the hematopoietic lineage (Ward-Kavanagh, Lin, Sedy, & Ware, 2016). HVEM serves both as a signaling receptor that is activated by its ligands and also as a ligand to activate BTLA and CD160 signaling (Cheung, Steinberg, et al., 2009; Sedy et al., 2013). Co-expression of HVEM with either BTLA or CD160 results in formation of a cell surface complex *in cis* that prevents binding of ligands *in trans* (Ware & Sedy, 2011). Interestingly, when HVEM was complexed to BTLA *in cis* it was unable to efficiently activate NF-κB signals in response to extracellular LIGHT, BTLA, CD160, or the HSV envelope protein gD (Cheung, Oborne, et al., 2009). It is not known whether BTLA or CD160 signaling is similarly reduced when these receptors are co-expressed with HVEM, although soluble HVEM proteins do not bind to BTLA-HVEM complexes (Sedy et al., 2017).

HVEM interactions with its ligands can be altered in pathologic settings, resulting in dysregulated immune responses. We previously found that human CMV evolved a paralog of HVEM (CMV ORF UL144) that specifically engages BTLA but does not bind CD160 (Sedy et al., 2013). This viral paralog of HVEM activates inhibitory signaling in T cells without engaging NK cell co-stimulation (Cheung et al., 2005). Indeed, we recently bioengineered a mutant form of HVEM containing four mutations that block LIGHT, LTα, and CD160 binding, and enhance BTLA affinity 10-fold (Sedy et al., 2017). These examples show how subtle molecular changes in the HVEM protein can dramatically alter its function.

3. Genetic lesions in *TNFRSF14* in lymphoid cancers

Leukemias and lymphomas are a diverse group of cancers arising from white blood cells in the bone marrow or the lymphatic system and that have been classified according to morphology, immunophenotype, genetic abnormalities, and clinical features (Swerdlow & International Agency for Research on Cancer & World Health Organization, 2008). A major basis for the classification of B cell neoplasms is their developmental stage indicating the cell of origin. In a number of recent studies researchers have used genomic analyses to characterize these cancers, identifying patterns of genetic lesions within different subsets of tumors that predict variant disease outcomes (Moffitt & Dave, 2017). The most prevalent form of lymphomas in the United States are non-Hodgkin B cell lymphomas, with Hodgkin lymphoma and lymphomas of T cell and NK cell origin comprising the remainder. B cells start their development in the bone marrow and then populate peripheral lymphoid organs as mature naive B cells. Normally, during B cell activation and progression through the GC reaction, B cells are selected for high affinity antibodies to ultimately generate long-lived memory B cells and plasma cells for optimal immune responses. Transformed B cell lymphomas and leukemias may arise at different stages in this pathway characteristic of distinct transcriptional profiles (Basso & Dalla-Favera, 2015; Rickert, 2013; Scott & Gascoyne, 2014; Shaffer, Young, & Staudt, 2012). B cell acute lymphocytic leukemias (B-ALL) arise from malignant pre-B cell clones in the bone marrow. While early B cells transiently express a pre-BCR that is maintained in some ALL cases, mature B cell neoplasms typically express and are dependent on isotype-specific signaling from the BCR. B cell chronic lymphocytic leukemias (B-CLL) arise from mature B cells in lymphoid tissues that have not begun the GC reactions or from cells that have exited the GC and are divided based on their immunoglobulin gene status into mutated or unmutated CLL, respectively. Several lymphomas derive from GCB including Burkitt lymphoma (BL), high-grade B cell lymphoma (HGBL), follicular lymphoma (FL), and diffuse large B cell lymphoma (DLBCL). The GCB and ABC subsets of DLBCL are transcriptionally distinct subtypes with ABC-DLBCL having characteristics of cells exiting the GC reaction (Schmitz et al., 2018). With the acquisition of additional genetic lesions indolent FL may convert to transformed FL (tFL) and progress to GCB-DLBCL or BL (Casulo, Burack, & Friedberg, 2015). Mucosa-associated lymphoid tissue (MALT) lymphoma arise from marginal zone B cells (MZB) that are localized to the marginal sinus of lymphoid tissues.

Genetic alterations in *TNFRSF14* were first confirmed in a cohort of FL in which the authors used deep sequencing methods to identify the specific mutations within the 1p36 genomic region that had previously been identified as frequently deleted or mutated in a range of cancers (Table 1) (Bjordahl et al., 2013; Cheung et al., 2010). Patients harboring these mutations had significantly worse outcomes compared to patients with wild-type *TNFRSF14* alleles. In a recent analysis of a large cohort of FL tumors, Pastore et al. quantified relative risk of disease in cohorts of patients with genetically profiled tumors (Pastore et al., 2015). Specifically, greater risk was predicted in patients harboring non-silent mutations in the genes *EP300*, *FOXO1*, *CREBBP*, and *CARD11*, and absence of mutations in *MEF2B*, *ARID1A* and *EZH2*, in addition to poor clinical score.

Table 1 HVEM network alterations in B cell malignancies.

Malignancy	Normal counterpart	Characteristic functional alterations	HVEM network alterations	References
B-ALL	Pre-B cell	↑BCR signaling (oncogene mimics), ↓Differentiation (*PAX5*—LOF)	*BTLA* (~7% mutated)	Liu et al. (2016), Mullighan et al. (2007), Muschen (2018), and Rickert (2013)
B-CLL	Mature B cell	↑BCR signaling, ↑NOTCH signaling, ↓Apoptosis (*miR-15a/16-1*—LOF), DNA repair (*ATM*—LOF)	*CD160* (~100% expressed)	Farren et al. (2011), Landau et al. (2015), Liu et al. (2010), Puente et al. (2015), Rodriguez-Vicente et al. (2017), and Ten Hacken, Gueize, and Wu (2017)
MALT lymphoma	Marginal zone B cell	↑NF-κB signaling, ↑BCR signaling (*H. pylori* infection)	*TNFRSF14* (~46% mutated in thyroid associated)	Bertoni, Rossi, and Zucca (2018) and Moody et al. (2018)
NMZL	Marginal zone B cell	↑NF-κB signaling, ↑NOTCH signaling	*TNFRSF14* (~17% mutated)	Bertoni et al. (2018) and Spina et al. (2016)
FL	GC B cell	↓Apoptosis (*BCL2*—GOF), epigenetic modification (*MLL2*—LOF), ↑mTOR signaling	*TNFRSF14* (~35% mutated)	Okosun et al. (2016) and Pastore et al. (2015)
GCB-DLBCL	GC B cell	↓Apoptosis (*BCL2*—GOF), Epigenetic modification (*EZH2*—GOF)	*TNFRSF14* (~40% mutated, ~80% in EZB subtype)	Pasqualucci and Dalla-Favera (2014), Reddy et al. (2017), and Schmitz et al. (2018)
ABC-DLBCL	GC B cell	↑NF-κB signaling, ↑BCR signaling, ↑Proliferation (*CDKN2A*—LOF), ↓Differentiation (*BLIMP1*—LOF)	*TNFRSF14* (~4% mutated)	Pasqualucci and Dalla-Favera (2014), Reddy et al. (2017), and Schmitz et al. (2018)
Hodgkin lymphoma	GC B cell	↑NF-κB signaling	*TNFRSF14* (~23% mutated)	Salipante et al. (2016)

↑, increased functionality; ↓, decreased functionality; GOF, gain of function; LOF, loss of function.

The rate of *TNFRSF14* mutations was increased in high risk patients, but this did not achieve statistical significance. It remains unclear whether *TNFRSF14* mutations may be more associated with particular genetic subtypes of FL. Analysis of tumor clonal diversity showed that mutations in *TNFRSF14* were predicted to arise late during lymphomagenesis and are potentially accelerator mutations that may provide growth advantages to cancers or may influence interactions with the immune system (Green et al., 2013).

Notably, *TNFRSF14* is repeatedly mutated in tFL (Pasqualucci et al., 2014). Among the 12 case pairs (FL and tFL) studied these mutations occur before transformation or are specifically acquired in tFL. Thus, the specific contribution to disease progression in tFL is likely but currently remains unclear. A better understanding of the specific tumor microenvironmental cues around HVEM signaling in FL is of utmost importance. Interestingly, in the primary cutaneous follicle center lymphoma subtype of FL, *TNFRSF14* mutations or 1p36 deletion were also more frequent than *BCL2* translocations or *EZH2* mutations, highlighting the central importance of HVEM in suppressing lymphoma development in this cancer subtype (Gango et al., 2018). *TNFRSF14* alterations were identified in cohorts of pediatric type FL tumors where mutations were associated with better prognosis in one study (Launay et al., 2012; Martin-Guerrero et al., 2013; Schmidt et al., 2016). Genomic analysis of pediatric FL revealed a high frequency of *TNFRSF14* mutations (up to 54%) similar to adult FL, frequent mutations in *MAP2K1* that promote ERK1/2 activation, absence of BCL2 or BCL6 translocations, and a low frequency of mutations for other genes typically associated with FL (*KMT2D*, *CREBBP*, *EP300*, *EZH2*) compared to adult FL (Louissaint et al., 2016; Schmidt et al., 2017). Interestingly, the majority of tumors harbored mutations in either *MAP2K1* or *TNFRSF14*, but not both, possibly indicating that mutation of either gene is sufficient to drive lymphomagenesis. A subset of tumors shared genetic lesions in both *MAP2K1* and *TNFRSF14* indicating that these are not mutually exclusive mutations. Together, these genomic analyses highlight prevalent *TNFRSF14* mutations in diverse forms of FL. Additionally, in some forms of (pediatric) FL *TNFRSF14* mutations may drive lymphomagenesis while in other (adult) forms *TNFRSF14* lesions may arise late during cancer development. Further work is required to determine how HVEM network signaling regulates diverse forms of FL.

In parallel to the genomic sequencing efforts in FL, sequencing of DLBCL tumor cohorts also revealed a high rate of *TNFRSF14* mutations (Table 1) (Lohr et al., 2012; Pasqualucci et al., 2011). Global genomic analyses revealed major cellular pathways repeatedly disrupted in DLBCL. These studies first showed that genetic alterations consist of mutations in genes that regulate gene accessibility (*EZH2, KMT2D, CREBBP, EP300*), survival (*BCL2*), immune escape (*B2M, CD58, CIITA*), and in NF-κB, JAK/STAT, and PI3K signaling pathways (*TNFAIP3, CARD11, MYD88, ITPKB, TRAF3*) (Pasqualucci & Dalla-Favera, 2014). A recent study correlating transcriptional changes in DLBCL with genomic changes in potential driver genes identified in a CRISPR screen showed that *TNFRSF14* alterations are associated with increased GCB-DLBCL phenotype, decreased infiltration of CD8$^+$ T cells and Treg, JAK/STAT inflammatory signaling, and increased cholesterol biosynthesis (Reddy et al., 2017). In a second landmark study in which DLBCL was categorized into genetic subtypes based on genomic signature, *TNFRSF14* alterations were highly associated with a subtype that contains frequent mutations in *EZH2* and *BCL2* (EZB subtype; *TNFRSF14* altered in 77.6% tumors), and that encompasses most tumors previously categorized as GCB-DLBCL (Schmitz et al., 2018). The EZB subtype was associated with the highest overall survival compared to other subtypes, consistent with the GCB-DLBCL category associated with greater survival compared to ABC-DLBCL. Additional frequent genetic lesions within the EZB subtype were found in genes that regulate several cellular processes potentially regulating oncogenesis including transcription and lineage commitment, cell survival, cellular signaling pathways, and immune editing. While HVEM has been described as a tumor suppressor, it is not entirely clear how it may participate in regulating these diverse cellular processes or how mutation of HVEM may provide a competitive advantage in DLBCL.

Recently, the prevalence of genetic deletions or mutations in *TNFRSF14* has become more widely appreciated in lymphomas and leukemias of diverse origin. In addition to FL and DLBCL, deletions of *TNFRSF14* were found in classical Hodgkin lymphoma, and a high prevalence of *TNFRSF14* mutations were found in certain types of mucosa-associated lymphoid tissue (MALT) lymphoma (Table 1) (Moody et al., 2018; Salipante et al., 2016). HVEM and LIGHT were found to be elevated in patients with cutaneous T cell lymphoma (CTCL), and also correlated with elevated levels of eosinophils, IL4, IL10, and soluble IL2R (Miyagaki et al., 2013). Higher HVEM expression was observed in patients with acute myeloid leukemia (AML) and was inversely correlated with pathogenic *FLT3-ITD* mutations, mutations in *NPM1*, and positively correlated with protective bi-allelic mutations in *CEBPA* (Lichtenegger et al., 2015). Importantly, elevated HVEM protein was associated with better outcome in one cohort with intermediate risk assessment.

A growing list of genes mutated in lymphoma has been validated in experimental systems and mouse models (Ramezani-Rad & Rickert, 2017). Within lymphoma, ablation of genes that regulate histone modification is frequently targeted (Morin et al., 2011). KMT2D is one methyltransferase that regulates transcriptional activity of genes in GCB including *TNFRSF14* (Ortega-Molina et al., 2015). Deletion of *Kmt2d* in mice was subsequently shown to increase B cell proliferation and frequency of GCB, consistent with its role as a tumor suppressor in lymphoma (Zhang, Dominguez-Sola, et al., 2015). BCL6 functions as a master regulator of GCB and a *BCL6* translocation to the immunoglobulin heavy chain promoter found in human DLBCL was modeled in mice (Cattoretti et al., 2005). These animals show increased GC formation and lymphoproliferative syndrome. Foxo1 is required in GCB for progression to the dark zone, and coordination of transcription with Bcl6 (Dominguez-Sola et al., 2015; Sander et al., 2015). In Burkitt lymphoma, Foxo1 mutations abolish its nuclear export, thus preventing its negative regulation and promoting proliferative signals (Kabrani et al., 2018). FOXO1 is directly regulated by PI3K signaling, which is frequently elevated in human lymphomas by PTEN deletions (Pfeifer et al., 2013). Deletion of both Pten and Inpp5d (Ship) phosphatases were modeled in animals leading to spontaneous tumor development (Miletic et al., 2010). Interestingly, animals deficient in both Pten and Inpp5d developed aggressive tumors of diverse histologic phenotypes, while in animals containing single deletions tumor development remained suppressed. Thus, partial dysregulation of PI3K signaling may broadly impact cellular signaling pathways (BCR, NF-κB, mTOR) to promote oncogenesis. In a landmark study examining the role of HVEM deletion mutants in FL, Boice et al. identified a role for HVEM as a tumor suppressor in a mouse model of lymphoma (Boice et al., 2016). Knockdown of HVEM in tumors resulted in a greatly accelerated onset of disease that was associated with increased activity of B cell signaling proteins associated with BCR activation. Additionally, HVEM knockdown showed increased activation of lymphoid stroma that were associated with elevated B cell expression of Tnf, Ltα, and Ltβ, and stromal expression of Cxcl13 and Ccl19, as well as increased frequencies of Il21-producing Tfh within tumors. Thus, HVEM plays multiple roles to activate inhibitory signals within this model system (Fig. 2). Interestingly, knockdown of BTLA in these tumors also accelerated lymphoma growth compared to controls. It remains to be seen whether HVEM activation of BTLA is the dominant inhibitory pathway in human lymphoma, and whether inhibitory signaling in B cells or Tfh is required to control tumor growth.

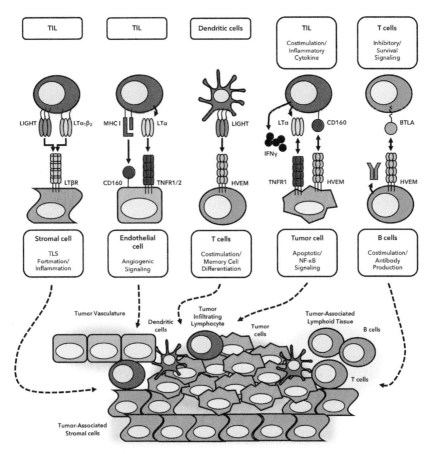

Fig. 2 HVEM network proteins regulate multiple cellular interactions within the tumor microenvironment. HVEM is expressed in TIL subsets and in many tumors within diverse types of cancer. HVEM activates CD160 in TIL to promote IFNγ production. HVEM activates BTLA in lymphocytes to activate both inhibitory signals and to promote cell survival. HVEM functions to co-stimulate lymphocyte activation, proliferation, effector function, and cell survival, but can also activate apoptotic signaling in some tumor cells similar to TNFR1. Within lymphoid structures T cells co-stimulate B cell activation, the GC reaction, and antibody production through HVEM, and B cells regulate Tfh activation through BTLA. LIGHT and $LT\alpha_1\beta_2$ stimulate TLS formation and inflammation through LTβR. Tumor vasculature expresses CD160 and TNFR1 receptors that can activate angiogenesis in response to ligand binding.

3.1 Immune microenvironment within lymphoma and influence of HVEM ligands

The wide expression of HVEM throughout hematopoietic cells indicates that it may play roles in regulating immune responses to lymphoma. Treg express increased HVEM protein compared to conventional $CD4^+$

T cells that may trigger inhibitory signaling in cells expressing BTLA (Tao, Wang, Murphy, Fraser, & Hancock, 2008). FOXP3$^+$ Treg were previously described as a component of tumor infiltrating immune cells in FL, with low frequencies of Treg associated with cancer that is refractory to treatment (Carreras et al., 2006). In contrast, high frequencies of PD-1$^+$ Treg in FL were associated with increased survival of patients with FL (Carreras et al., 2009). In FL, PD-1$^+$ Tfh cells that express IL4 were found to reside in close proximity to B cells, driving STAT6 signaling (Pangault et al., 2010). In lymphoma patients, HVEM activation of BTLA receptors was found to negatively regulate tumor-reactive TCR Vγ9Vδ2$^+$ T cells (Gertner-Dardenne et al., 2013; Sedy et al., 2014). However, BTLA has been shown to activate both inhibitory and survival signaling, and its function in tumor-reactive lymphocytes may be context specific (Murphy & Murphy, 2010). Together, it is clear that lesions in the *TNFRSF14* gene are widely present in lymphomas of B cell origin, and its expression in non-B cells present in the tumor microenvironment may also influence cancer development.

4. Genetic lesions in *TNFRSF14* in non-lymphoid cancers

Fewer studies have linked specific genetic changes in *TNFRSF14* to disease in non-lymphoid cancers. However, HVEM expression is often induced in many tumors and is associated with reduced survival in these settings. Notably, elevated expression of HVEM was observed in many melanoma cell lines and in human tumors with a corresponding increase in BTLA expression in circulating melanoma-specific CD8$^+$ T cells (Derre et al., 2010; Fourcade et al., 2012). Antagonist antibodies that blocked BTLA binding to HVEM could rescue T cell proliferation *in vitro*, while CpG vaccination could reduce BTLA expression and expand melanoma-specific T cells in human patients. These data highlight how the inhibitory function of BTLA in anti-tumor responses can be removed with checkpoint blockade therapy.

A genetic polymorphism (rs2234167) in *TNFRSF14* that results in a V241I mutation in the cytoplasmic region of HVEM was found to be associated with reduced risk for breast cancer in a cohort of Chinese women (Li et al., 2013). Additionally, elevated HVEM expression was found to be associated with reduced survival in breast cancer, with greatest risk in those patients having a low frequency of TIL (Tsang et al., 2017). HVEM was also found to be significantly upregulated in ovarian cancer, and that silencing HVEM in ovarian cancer tumors stimulated greater T cell activation compared to control cells (Fang, Ye, Zhang, He, & Zhu, 2017; Zhang, Ye, Han, He, & Zhu, 2016). HVEM expression in esophageal squamous cell carcinoma was found to be associated with decreased overall survival, and gene silencing of HVEM decreased tumor proliferation in a mouse model (Migita et al., 2014). In gastric cancer elevated levels of both BTLA and HVEM proteins were associated with worse prognosis (Lan et al., 2017). HVEM was shown to be expressed in human colorectal cancer with no expression in normal colon tissue, and HVEM expression was inversely correlated with the presence of TIL, as well as overall survival of patients (Inoue et al., 2015). In one study of hepatocellular carcinoma, HVEM expression was also shown to be expressed in tumors compared to low expression in normal liver tissue, with expression inversely correlating with the presence of TIL and expression of *PRF1*, *GZMB*, and *IFNG*, and again was correlated with reduced survival (Hokuto et al., 2015). In a second analysis of hepatocellular carcinoma patients, high expression of HVEM, as well as of PD-L1, Galectin 9, IDO, and the presence of TIL were associated with increased survival (Sideras, Biermann, Verheij, et al., 2017). LIGHT was found to be elevated in inflamed livers from patients with nonalcoholic fatty liver disease, and thus may stimulate inflammation in pre-cancerous livers through HVEM (Otterdal et al., 2015; Taniguchi & Karin, 2018). In a parallel study of pancreatic and ampullary cancer, it was found the high expression of either HVEM, PD-L1, Galectin 9, IDO, HLA-G, along with a high ratio of $CD8^+$ T cells to $FOXP3^+$ Treg was associated with increased survival (Sideras, Biermann, Yap, et al., 2017).

Together these data illustrate an emerging picture in which HVEM expression is associated with diverse outcomes in cancer (Fig. 2). In B cell lymphomas growing within a GC microenvironment HVEM appears to function as a tumor suppressor, acting in part to activate BTLA inhibitory signals in Tfh cells, and potentially in B cell lymphoma themselves. In other solid tumors patient survival is closely linked with the frequency of TIL. HVEM expression within these tumors may function to inhibit lymphocyte responses through activation of BTLA inhibitory receptors. Alternatively, HVEM may bind CD160 receptors to activate anti-tumor responses. Finally, the role of HVEM-BTLA or HVEM-CD160 complexes *in cis* in tumors that may express these receptors such as in B cell lymphoma or AML is not understood. B cell lymphoma lacking HVEM may show increased BTLA triggered inhibitory signals *via* SHP1/2 activation or PI3K survival signaling. The function of HVEM in cancer may depend on the nature of the immune microenvironment present in different tumors (Thorsson et al., 2018).

5. HVEM functions within the tumor microenvironment

The distribution of point mutations in the coding region of HVEM provides some insight into the selective pressures driving *TNFRSF14* genomic alterations in cancer (Fig. 3). In lymphoma, the *TNFRSF14* mutations were previously observed to be enriched in the extracellular domain,

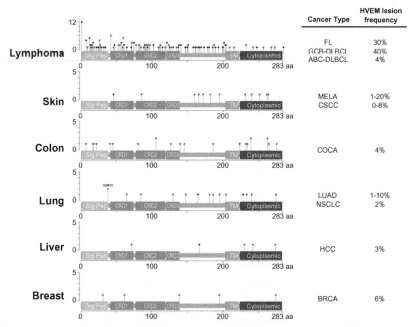

Fig. 3 Somatic mutations within the protein sequence of human HVEM. Mutations described in human HVEM are shown for cancers of lymphocytes, skin, colon, lung, liver, and breast tissue. Lollipop plots depict the HVEM protein sequence including signal peptide (yellow), CRD1–3 (green), transmembrane domain (TM, light blue), and cytoplasmic domain (red), and the positions and frequency of translational start mutations (purple), missense mutations (green), in-frame insertion/deletion (brown), or truncating mutation (black). The percentage of tumors containing *TNFRSF14* lesions in the corresponding tumor dataset is indicated. Mutations were identified in publicly available databases based upon data generated in part by the TCGA Research Network: http://cancergenome.nih.gov/, nnCOSMIC, and the ICGC (Forbes et al., 2017; Gao et al., 2013; Pastore et al., 2015; Schmitz et al., 2018; Zhang et al., 2011).

indicating that ligand interactions may be preferentially targeted in these cancers (Bjordahl et al., 2013). While the functional impact of these mutations remains to be determined, it has been suggested that mutations in the extracellular domain of HVEM may prevent interactions with its ligands LIGHT, LTα, BTLA, and CD160, thus preventing HVEM activation and signaling. Alternatively, these mutations may prevent HVEM bidirectional signaling through its counterreceptors BTLA and CD160, which are activated upon HVEM binding. Several point mutations have also been observed in the cytoplasmic domain of HVEM. In a recent survey of mutations in *TNFRSF14*, we observed a greater frequency of mutations located within the cytoplasmic domain of HVEM in cutaneous melanoma and in colorectal adenocarcinoma. We predict mutations in the cytoplasmic domain of HVEM to prevent either HVEM signaling entirely or to alter the nature of HVEM signaling by influencing the kinetics of adaptor protein recruitment.

5.1 LTα and LIGHT activate pleiotropic anti-tumor functions

The activity of LTα has predominantly been attributed to TNFR1 and TNFR2 activation (Fig. 2). However, in humans its weak affinity for HVEM indicates potential actions through this receptor. Soluble LTα protein forms a homotrimer that functions as a ligand for TNFR1 and TNFR2 receptors to activate apoptotic signaling in tumors. LTα additionally forms a heterotrimeric complex with cell membrane-expressed LTβ (LTα$_1$LTβ$_2$) that binds LTβR receptors to activate lymphorganogenesis (Ware, 2005). The homotrimeric form of LTα itself was shown to promote TLS within tumors, resulting in increased T cell infiltration and activation (Schrama et al., 2001). Both LTα and LIGHT may participate in development of TLS in autoimmunity and in cancer (reviewed in Tang, Zhu, Qiao, & Fu, 2017). Mutations within non-coding regions of the *LTA* gene have been identified in a number of cancers, and in meta-analyses SNPs were shown to be associated with susceptibility to disease in inflammation-related cancers, gastric cancer, non-Hodgkin lymphoma, and leukemia (Cao, Liu, Lou, & Liu, 2014; Gong et al., 2017; Huang et al., 2013; Xu, Shi, Zhang, Zhang, & Wang, 2013; Yang et al., 2013; Yu et al., 2014). In classical Hodgkin lymphoma, Reed-Steinberg tumors were shown to secrete LTα that could activate endothelial cells to express hyaluronan that in turn recruits CD4$^+$ CD45RA$^+$ naïve T cells, while in CTCL, lymphoma cells were also shown to express LTα that stimulated IL6 and angiogenesis in tumors

(Fhu et al., 2014; Lauenborg et al., 2015). In this context it was not clear whether LTα is acting through TNFR1, TNFR2, or HVEM receptors. LTα as well as IL12 and TGFβ could potentiate NK activity when cultured with DC to promote cytokine production and cytolysis (Sarhan et al., 2015).

LIGHT was first shown to stimulate anti-tumor responses in a mouse model of melanoma (Yu et al., 2004). Tumors expressing a non-cleavable form of LIGHT could activate T cell responses through HVEM and local inflammation through LTβR signaling (Fig. 2). Importantly, activated T cells could respond to tumors inoculated at a distal site and which did not express LIGHT. LTβR was subsequently shown to be necessary for LIGHT-mediated anti-tumor activity, resulting in increased $CD8^+$ T cell activation and inflammatory production that cooperated with checkpoint blockade (Tang et al., 2016). LIGHT was additionally shown to activate HVEM in NK cells to stimulate IFNγ that was then required for optimal cytotoxic $CD8^+$ T cell activity (Fan et al., 2006). In addition to NK cells, LIGHT has been shown to co-stimulate HVEM signaling in a number of cell types to drive primary anti-tumor responses, including in T cells, neutrophils, macrophages, and other effector cells within tumors to activate cytolysis, cytokine secretion, and inflammatory signaling (Murphy, Nelson, & Sedy, 2006; Sedy et al., 2014). LIGHT activation of HVEM additionally has been shown to induce apoptosis of several tumor cell lines directly. In NK cells, LIGHT expression is stimulated by CD16, NK receptors, or cytokine receptor signaling, and was found to stimulate DC activation through HVEM (Cheung, Steinberg, et al., 2009; Holmes et al., 2014). Importantly, in NK cells, tumor cells induced LIGHT expression to the greatest extent in those licensed cells that expressed a greater diversity of inhibitory NK receptors. LIGHT was also shown to be induced in cells activated through the innate receptor AXL and PI3K signaling (Lee et al., 2017). LIGHT was associated with increased numbers of TIL within colorectal liver metastases, indicating that it regulates both primary and secondary anti-tumor responses (Maker et al., 2015). Recently, LIGHT activation of HVEM and LTβR has been shown to stimulate fibrosis in inflammatory mouse models in coordination with TSLP and induced expression of the extracellular matrix protein Periostin (Herro & Croft, 2016). The role of fibrosis in cancer is context dependent, and it remains to be determined whether in particular cancers HVEM and LTβR activated fibrosis limits anti-tumor responses through support of stroma, or whether this limits tumor angiogenesis and metastasis. In multiple myeloma patients, LIGHT was reported to be highly expressed by $CD14^+$ monocytes, $CD8^+$ T cells, and neutrophils, where it may contribute to osteoclastogenesis and cancer progression (Brunetti et al., 2014). In primates, LIGHT-mediated activation of its receptors is counterbalanced by the expression of DcR3 which competes with HVEM and LTβR for LIGHT binding, neutralizing its signaling (Liu et al., 2014; Ward-Kavanagh et al., 2016). Together, these reports illustrate pleiotropic functions of LIGHT in anti-tumor immune responses, and how its role in specific cancers depends on the tumor context.

5.2 BTLA identifies exhausted T cells

BTLA functions to inhibit activation of antigen receptor stimulation in T cells and B cells upon HVEM binding through the recruitment of the phosphatases SHP1 and 2, similar to the checkpoint proteins CTLA-4 and PD-1 (Fig. 2) (Ward-Kavanagh et al., 2016). Recently, BTLA has been shown to inhibit inflammatory cytokine production activated by Toll-like receptor signaling in DC, and by IL7 signaling in γδ T cells (Bekiaris, Sedy, Macauley, Rhode-Kurnow, & Ware, 2013; Kobayashi et al., 2013). BTLA may additionally recruit GRB2 and the p85 subunit of PI3K to activate the AKT pathway and downstream survival signals (Gavrieli & Murphy, 2006; Murphy & Murphy, 2010). As previously described, BTLA blockade could enhance circulating $CD8^+$ T cell anti-melanoma responses (Derre et al., 2010; Fourcade et al., 2012). However, in a recent study of tumor infiltrating $CD8^+$ T cells in metastatic melanoma, Haymaker et al. demonstrated that $BTLA^+$ cells were more responsive and showed greater longevity *in vivo* compared to $BTLA^-$ cells (Haymaker et al., 2015). Interestingly, activation of these $CD8^+$ cells by HVEM increased survival signaling through Akt, although it was not clear which HVEM receptor is being activated in this context. In a subsequent study of human melanoma, increased expression of CD8A and BTLA correlated with increased survival that was associated with greater tumoricidal activity (Ritthipichai et al., 2017). However, lack of BTLA did not specifically affect $CD8^+$ T cells cytolytic function. In contrast, BTLA was required for efficient memory responses *in vivo*. Additionally, in a study of melanoma patients treated with adoptive cell therapy, BTLA expression in TIL was associated with a less differentiated phenotype and greater responsiveness to IL2 and TCR stimulation as well as response to therapy, but not following anti-CTLA-4 blockade treatment (Forget et al., 2018; Haymaker et al., 2015). Interestingly, HVEM activation of BTLA-expressing cells both inhibited proliferation and expression of IFNγ, and activated survival signaling through AKT.

Thus, in the context of melanoma, BTLA may regulate generation of effector cells, rather than effector function specifically.

Activation of BTLA by HVEM in lymphoma-reactive human TCR Vγ9$^+$Vδ2$^+$ T cells resulted in inhibition of proliferation and partial cell-cycle arrest that could be rescued by blocking BTLA ligation (Gertner-Dardenne et al., 2013). Interestingly, IL12 and IL18 stimulation could enhance anti-tumor responses of TCR Vγ9$^+$Vδ2$^+$ T cells while decreasing surface expression of BTLA (Domae, Hirai, Ikeo, Goda, & Shimizu, 2017). Thus, one of the selective pressures that shape the lymphoma microenvironment is to maintain HVEM binding to BTLA binding to regulate innate $\gamma\delta$ T cell anti-tumor responses.

In gastric cancer high BTLA expression was associated with reduced survival, potentially linked to its expression in CD40$^+$ MDSC (Ding et al., 2015; Feng et al., 2015). DC within tumors of urothelial cancer patients were found to express high levels of BTLA and TIM-3 that suppressed the production of IL12 p70, IL1β, and TNF (Chevalier et al., 2017). In gall bladder cancer high levels of BTLA in tumor infiltrating CD8$^+$ and CD4$^+$ T cells was associated with significantly reduced patient survival (Oguro et al., 2015). In human hepatocellular carcinoma (HCC) BTLA$^+$ CD4$^+$ T cells showed diminished functionality compared to BTLA$^-$PD-1$^-$ or BTLA$^-$PD-1$^+$ cells, with the least functionality observed in PD-1$^+$BTLA$^+$ cells that were associated with late stages of disease (Zhao, Huang, He, Gao, & Kuang, 2016). In non-small cell lung cancer (NSCLC) high BTLA expression in CD8$^+$ T cells was observed in patients with late stage disease, and correlated with expression of CTLA-4, PD-1, TIM-3, and LAG-3 (Thommen et al., 2015). Similarly, in DLBCL patients, BTLA expression in T cells identified cells co-expressing PD-1, LAG-3, and TIM-3 that showed a less differentiated phenotype, and less functionality of BTLA$^-$ T cells (Quan et al., 2018).

Thus, BTLA co-expression with other inhibitory receptors may define cells that receive substantial inhibitory signals, termed "exhausted" T cells. T cell activation and differentiation alone promote increased inhibitory receptor expression, requiring careful evaluation of cell phenotypes (Legat, Speiser, Pircher, Zehn, & Fuertes Marraco, 2013). The factors that induce development of exhausted cells remains unclear, although recent efforts indicate that this cell fate may be the result of a specific transcriptional program (Attanasio & Wherry, 2016; Doering et al., 2012). In a mouse model of melanoma, CD4$^+$ T cell expression of BTLA mRNA as well as PD-1 and TIM-3 transcripts was found to be controlled by miR-28, which decreases protein expression of these receptors and their inhibitory activity (Li et al., 2016).

In colorectal cancer, disease risk was associated with BTLA intron polymorphisms and in renal cell carcinoma disease risk was associated with polymorphisms in the 3′ untranslated region (Ge et al., 2015; Partyka et al., 2016).

As previously described, BTLA expression in tumors originating in the hematopoietic lineage can influence tumor burden in animal models and in cancer patients. In human FL, BTLA was found to be silenced in half of all tumors, and that BTLA silencing was inversely correlated with the presence of HVEM lesions (Boice et al., 2016). Thus, in the lymphoma microenvironment a selective pressure may be to block HVEM-BTLA interactions between tumor cells. Within B-ALL recurrent deletions were observed in the locus containing BTLA and CD200 resulting in loss of both genes that were associated with reduced survival (Ghazavi et al., 2015). A SNP in BTLA was found to be associated with risk for CLL as well as reduced mRNA expression in T cells but not B cells that may decrease the threshold of B cell activation and potentially oncogenic signaling (Karabon et al., 2016). Interestingly, analysis of CLL patients treated with BTK or ITK inhibitors showed a significant reduction of CLL cells expressing both BTLA and CD200, as well as a significant reduction in IL10-producing $CD19^+CD5^+$ CLL cells that share the phenotype of Bregs (Long et al., 2017).

5.3 CD160 co-stimulates anti-tumor IFNγ

CD160 is a cell surface receptor expressed in NK cells, $CD8^+$ T cells, $TCR\alpha\beta^+$ and $TCR\gamma\delta^+$ intraepithelial lymphocytes in the intestine, and skin resident $CD4^+CD8\alpha\alpha^+$ cytotoxic T cells that stimulates effector function following engagement by its ligands (McDonald, Jabri, & Bendelac, 2018; Sako et al., 2014; Ward-Kavanagh et al., 2016). The identified ligands for CD160 receptors include MHC class I proteins and HVEM. CD160 engagement by HVEM in NK cells promotes co-stimulation of ERK1/2 and AKT activation and production of IFNγ (Fig. 2) (Sedy et al., 2013; Ward-Kavanagh et al., 2016). In contrast, in $CD4^+$ T cells HVEM-mediated activation of CD160 induces inhibitory signaling (Cai et al., 2008). Thus, CD160 function may be dependent on its cellular context.

CD160 co-expression with inhibitory receptors in tumor-reactive T cells has been characterized as an exhausted phenotype (Attanasio & Wherry, 2016). In CLL and AML CD160 is co-expressed with 2B4 and PD-1 in T cells, although expression of these receptors did not impact

T cell functional responses toward viral antigens (Riches et al., 2013; Schnorfeil et al., 2015; te Raa et al., 2014). In multiple myeloma patients, CD160 is co-expressed with PD-1, CTLA-4, and 2B4 in bone marrow infiltrating $CD8^+$ T cells and correlates with reduced proliferative capacity (Zelle-Rieser et al., 2016). However, in T cells recovered from adoptive cell therapy CD160 expression correlated with reduced functionality in a ligand-independent fashion (Abate-Daga et al., 2013).

CLL B cells themselves frequently express CD160 that activates PI3K-dependent oncogenic signals in these tumors, and CD160 expression has been proposed as a means to detect minimal residual disease (MRD) in patients with remission (Farren et al., 2015, 2011; Lesesve et al., 2015; Liu et al., 2010; Zhang, Chen, Chen, Liu, & Qu, 2015). Interestingly, CD160 is absent in many mycosis fungoides T cell lymphomas in the skin or in circulation, potentially due to the stage of disease (Sako et al., 2014). CD160 expression was also identified within endothelial cells in newly formed vasculature, including that of tumors, but not in blood vessels of normal tissues (Chabot et al., 2011; Fons et al., 2006).

The specific function of CD160 was tested *in vivo* using a genetic model of *Cd160*-deficiency challenged with B16 melanoma cells (Tu et al., 2015). *Cd160*-deficient NK cells showed impaired tumor induced IFNγ production and were less efficient in clearing tumors compared to wild-type cells. IFNγ has pleiotropic effects in tumors, including inducing expression of HVEM, FAS, ICAM1, PD-L1, and MHC class I and II proteins (Aquino-Lopez, Senyukov, Vlasic, Kleinerman, & Lee, 2017). Thus, IFNγ that is activated by CD160 in NK cells further amplifies CD160 signaling through increased HVEM expression in tumors that may result in increased tumor lysis.

5.4 Intracellular signaling pathways activated by HVEM

HVEM is expressed widely throughout the hematopoietic system, epidermal and endothelial cells, and at low levels in neurons (Ward-Kavanagh et al., 2016). The function of HVEM was first characterized in T cells where it was identified as a co-receptor for T cell receptor activation (Murphy et al., 2006). HVEM was first shown to be activated by its ligands LIGHT and LTα, and then later by the ligands BTLA and CD160, resulting in activation of NF-κB signaling through recruitment and stabilization of cytoplasmic TRAF2 (Bechill & Muller, 2014; Murphy & Murphy, 2010). HVEM may also recruit additional TRAF proteins including TRAF3 to regulate both canonical and non-canonical NF-κB signaling (Ward-Kavanagh et al., 2016). However, the activation of the alternative NF-κB pathway is greatly reduced compared to other TNFR family members such as LTβR that drive robust p100 processing and RELB nuclear translocation (Cheung, Steinberg, et al., 2009). A structure of TRAF2 bound to the cytoplasmic domain of HVEM has not been solved. However, the cytoplasmic domain of human HVEM is predicted to bind TRAF2 using a peptide sequence located at the C-terminus of the protein that aligns with the major TRAF binding motif (HVEM 266–276: TV<u>AVEETI</u>PSF; core residues underlined) (Ye, Park, Kreishman, Kieff, & Wu, 1999). In T cells, HVEM has been linked to PI3K and AKT signaling and activation of survival signals (Flynn et al., 2013; Soroosh et al., 2011; Steinberg et al., 2013). Indeed, many of the TNFR family members including HVEM have been proposed to activate PI3K signaling through as of yet undescribed mechanisms that may involve ligand induced receptor clustering and association with DIMs (Fig. 1) (So & Croft, 2013). In an animal model of acute/effacing enteropathogenic *E. coli* infection, HVEM was shown to activate NIK-dependent Stat3 phosphorylation to induce Reg3γ production in epithelial cells in response to Cd160 binding in the intestine (Shui et al., 2012). However, the molecular mechanism of Stat3 activation has not been specifically determined. LIGHT activation of cells has been shown to induce MAPK activation resulting in phosphorylation of p38, JNK, and ERK proteins, however the receptor initiating these signals is often unclear due to the shared binding of LIGHT for HVEM and LTβR (Bechill & Muller, 2014; Murphy et al., 2006).

While missense mutations have been identified in the cytoplasmic domain of human HVEM in several cancers, it remains unclear how these mutations may impact HVEM signaling or tumor fitness. Several missense mutations have been identified in the cytoplasmic domain of HVEM in human B cell lymphoma including G232S (COSMIC), Q242H (ICGC), E256Q (TCGA), V267M (COSMIC), T272I (COSMIC), and similar mutations are present in other cancers (Fig. 3) (Forbes et al., 2017; Gao et al., 2013; Zhang et al., 2011). It remains unclear whether these mutations impact HVEM recruitment of TRAF proteins and downstream NF-κB signaling, or the activation of alternate signaling pathways that may influence oncogenesis. In DLBCL, genes encoding NF-κB signaling proteins are frequently mutated, resulting in dysregulated downstream gene activation (Schmitz et al., 2018). For example, in lymphoma the TRAF3 gene is frequently mutated, preventing its normal function as part of the ubiquitin ligase complex that is also composed of TRAF2, BIRC2, and BIRC3, and that normally targets NIK for proteosomal degradation (Sanjo, Zajonc, Braden, Norris, & Ware, 2010; Vallabhapurapu et al., 2008). In a mouse model, enforced NIK expression with a Bcl6 transgene promoted B cell hyperplasia (Zhang, Calado, et al., 2015). However, TRAF2 and TRAF3 have been shown to have diverse functions in different cancer settings. TRAF2 was recently shown to function as a ubiquitin ligase that targets the mTOR complex component GβL whose function is opposed by the deubiquitinase OTUD7B (Wang et al., 2017). Defective TRAF2 ubiquitination of GβL is proposed to stabilize its interaction with the mTORC2 component SIN1, thus promoting oncogenic mTORC2 activity (Senft, Qi, & Ronai, 2018).

LIGHT was suggested to directly render leukemia and lymphoma cells more sensitive to FAS induced apoptosis *via* HVEM, and that mutations in the extracellular domain would prevent HVEM signaling (Costello et al., 2003; Lohr et al., 2012). In a B-CLL cell line, HVEM signaling induced expression of pro-apoptotic genes including FASL, and triggered cleavage of caspase-3 (Pasero et al., 2009). In contrast, LIGHT activation of LTβR was previously shown to induce apoptosis in a variety of cancer cell lines in a TRAF3-dependent fashion and despite NF-κB activation (Bechill & Muller, 2014; Hu et al., 2013; Rooney et al., 2000). Co-expression of HVEM with LTβR may thus prevent LIGHT-induced apoptosis in some cellular contexts. It is unclear whether HVEM regulates TRAF2 and TRAF3 protein activation in a different manner than LTβR, or whether HVEM activates anti-apoptotic signaling through some other mechanism.

HVEM expression was identified in HSC, but not in bone marrow-derived mesenchymal stem cells, and LIGHT could stimulate HSC expression of GM-CSF, GM-CSF receptor, downstream PU.1 expression, and myeloid differentiation (Chen et al., 2018; Heo et al., 2016). Thus, HVEM may regulate stem cell functions within the hematopoietic lineage that may participate in oncogenic signaling. LIGHT has also been shown to activate HVEM and LTβR-mediated signaling to activate inflammatory signaling through coordination of the cytokine TSLP and production of Periostin to induce fibrosis (Herro, Antunes, Aguilera, Tamada, & Croft, 2015; Herro, Da Silva Antunes, Aguilera, Tamada, & Croft, 2015; Herro et al., 2018). Together, these findings illustrate diverse functions of HVEM across cancers. Molecular analysis of cell-extrinsic *versus* cell-intrinsic interactions in these tumors will identify how HVEM functions to regulate oncogenesis *in vivo*.

In addition to its function in tumors, HVEM was shown to co-stimulate the activity of a wide variety of lymphoid and myeloid cells including T cells, B cells, NK cells, DC, neutrophils, and monocytes largely through experiments using the TNF ligand LIGHT. BTLA and CD160 were also shown to activate co-stimulatory pathways in T cells, with the limitation that these ligands must compete BTLA or CD160 on HVEM expressing cells (Ware & Sedy, 2011). The ligand SALM5 binds to HVEM in the same portion of the extracellular domain as BTLA and CD160 and is activated by HVEM binding (Zhu et al., 2016). However, it is unclear whether SALM5 can act as an activating ligand for HVEM in the CNS, and how this interaction functions in the context of neuronal cancers.

6. Therapeutic targeting of the HVEM network in lymphoma and other tumors

Checkpoint blockade therapy designed to disrupt CTLA-4 and PD-1 signaling provides durable responses for a subset of patients and cancers. However, a substantial proportion of patients do not respond to these immunotherapies, necessitating the development of novel cancer treatments including additional immunotherapies targeting a broader spectrum of inhibitory receptors alone and in combination with established treatments. Many such drugs are currently being evaluated in experimental models and in clinical trials to increase success rates in treatable cancers and in currently intractable cancers (Carter & Lazar, 2018). The HVEM network of receptors and ligands has long been viewed as potentially druggable in cancer, but only recently have HVEM network-targeting immunotherapies entered into pharma pipelines (Table 2) (Croft, Benedict, & Ware, 2013; Pasero & Olive, 2013). In contrast, multiple efforts have been made to block inflammatory functions of HVEM network proteins in autoimmune disease.

6.1 LTα

Initial studies using soluble LTα to treat melanoma in a mouse model *in vivo* demonstrated that this protein could reduce tumor burden and increase survival that was associated with formation of TLS within tumors (Schrama et al., 2001). *De novo* mutant proteins have been developed to increase the affinity of LTα for TNFR1, and its apoptotic potential, and are currently being evaluated in clinical trials (Morishige et al., 2013; Wang et al., 2016). It was not reported whether these proteins alter binding to HVEM. LTα and TNF-mediated apoptosis can also be activated through inhibition of BIRC2 and BIRC3 proteins using SMAC mimetics. A recent effort to screen small molecules yielded potent compounds that could potentiate lytic activity in a prostate tumor cell line (Welsh et al., 2016).

Table 2 Therapeutic targeting of the HVEM network in human disease.

Gene	Therapy	Indication	Status/References
HVEM	HVEM agonist[a]	Immuno-Oncology	Preclinical (Tang, Pearce, O'Donnell-Tormey, & Hubbard-Lucey, 2018)
BTLA	Antagonist anti-BTLA[a]	Immuno-Oncology	Preclinical
	Agonist anti-BTLA[a]	Inflammatory disease	Preclinical
CD160	Agonist anti-CD160[a]	Immuno-Oncology	Preclinical (Tang et al., 2018)
	Soluble CD160 biomarker	Ischemic retinopathy	NCT03680794
LIGHT	Antagonist anti-LIGHT (MDGN-002)[a]	Inflammatory bowel disease	NCT03169894 (Croft & Siegel, 2017)
LTα	Etanrecept (TNFR2-Fc, neutralizing TNF, LTα)	Rheumatoid arthritis, Psoriatic arthritis, Psoriasis, Juvenile idiopathic arthritis, Ankylosing spondylitis	Approved (Enbrel®) (Croft & Siegel, 2017)
	Pateclizumab (anti-LTα)	Rheumatoid arthritis	NCT01225393 (Kennedy et al., 2014)
LTβR	Baminrecept (LTβR-Fc, neutralizing LIGHT, LTαβ)	Rheumatoid arthritis (failed to meet endpoint), Sjögren's syndrome (expired drug), Secondary progressive multiple sclerosis (withdrawn)	NCT00523328, NCT00664573, NCT00664716, NCT01164384, NCT01181089, NCT01552681 (Bienkowska et al., 2014)

[a]Antibodies and biologics directed against cell surface receptors that block ligand interactions are defined as antagonists, while those that activate receptor signaling are defined as agonists.

6.2 LIGHT

The potent inflammatory activity of LIGHT has encouraged interest in potential therapeutic applications including delivery of LIGHT proteins to tumor sites. Forced expression of LIGHT in prostate tumors promoted an increase of TIL and a concurrent decrease in Treg activity (Yan et al., 2015). LIGHT in this setting was presumed to activate T cells directly through HVEM expressed in T cells, although it may be possible that LTβR activated inflammatory signaling is activated in stromal cells. Induced expression of LIGHT in a colorectal carcinoma cell line could co-stimulate T cell activation, increased frequency of TIL, and decreased tumor burden *in vivo* (Qiao et al., 2017). Soluble LIGHT proteins targeting vessels or LIGHT expressing macrophages were effective in normalizing tumor vasculature to promote tumor perfusion in a TGFβ- and ROCK1-dependent fashion, increasing survival in a model of pancreatic neuroendocrine and orthotopic breast cancers (Johansson-Percival et al., 2015). Vasculature-targeted LIGHT proteins could similarly normalize blood vessels as well as induce T cell infiltration and high endothelial venule formation in a model of glioblastoma *in vivo* (He et al., 2018). Vasculature-targeted LIGHT proteins induced TLS that were dependent on T cells and macrophages, activated T cell effector and memory responses, facilitated tumor infiltration and lysis, and enhanced checkpoint blockade and anti-tumor therapy (Johansson-Percival et al., 2017). However, as these studies were carried out in a mouse experimental system that lacks LIGHT-neutralizing DcR3 proteins, the feasibility of this approach in humans remains to be validated.

6.3 BTLA

Blockade of HVEM-BTLA inhibitory signaling using antagonistic antibodies has been proposed as a potential therapeutic for cancer based on promising *in vitro* studies and in animal models (Derre et al., 2010; Fourcade et al., 2012). In an alternative approach, soluble BTLA proteins expressed *via* adeno-associated virus promoted rejection of melanoma in a mouse model (Han et al., 2014). In yet another approach, HVEM peptides were used to block BTLA binding to purified HVEM proteins, though not in a cellular context (Spodzieja et al., 2017). In FL, genetic abnormalities in *TNFRSF14* were associated with increased T cell activation following allogeneic hematopoietic stem cell transplantation (Kotsiou et al., 2016). These data indicate that lymphoma-expressed HVEM may activate BTLA receptors in donor T cells to limit GVHD and anti-tumor responses. While PD-1 blockade alone *in vitro* uniquely enhanced T cell activation compared to other inhibitory receptors, additional blockade of BTLA significantly enhanced T cell proliferation and cytokine production (Stecher et al., 2017). In the PyMT mouse model of breast cancer, anti-BTLA blockade could reduce tumor burden that was associated with increased infiltration of BTLA$^+$ NKT, and in human breast cancer increased expression of the NKT cell lineage transcription factor ZBTB16 was associated with increased survival

(Sekar et al., 2018). CAR-T cell therapy is a promising novel means to activate antigen-specific anti-tumor responses that were recently used to additionally supply soluble HVEM proteins in a mouse lymphoma model (Boice et al., 2016). However, the tumor suppressor function of HVEM in DLBCL to activate BTLA inhibitory signaling is likely unique to this cancer.

6.4 CD160

Co-stimulatory functions of CD160 in NK cells indicated a potential use for agonistic antibodies in mobilizing innate anti-tumor responses (Sedy et al., 2014). Additionally, endothelial expression of CD160 during neoangiogenesis revealed additional potential oncotherapeutic targets (Fons et al., 2006). Interestingly, Chabot et al. found that CD160-specific agonist antibodies could reduce tumor burden in animals inoculated with B16 melanoma (Chabot et al., 2011). Anti-CD160 reduced tumor blood vessel number, size, and branching, likely contributing to reduced tumor volume. However, activation of NK cells was not evaluated in this model, and it remains unclear to what extent anti-CD160 regulation of tumor growth is dependent on vasculature or innate cell compartments. Cellular therapy using NK cells is a promising potential anti-tumor treatment due to the innate ability of these cells to recognize many cancers. However, development of NK cell therapy continues to face many challenges in the efficiency and purity of cellular expansion from donors, and the quality of cytotoxic function (Klingemann, 2013). Expression of CD160 within CAR-T cells may promote co-stimulation, IFNγ production, and memory cell formation from these engineered cells, thus improving on existing cellular therapies.

7. Concluding remarks

It is clear that HVEM network proteins play critical roles in regulating immune responses to cancer, while emerging data suggest that HVEM itself may regulate the oncogenic process. The pleiotropic functions of this network in diverse cellular settings make it difficult to identify dominant cellular pathways driving tumor selection in diverse cancer settings, and how best to target this network using oncotherapy. Nevertheless, certain patterns have emerged in past analyses. Much of the data confirm inflammatory functions for LIGHT in cancer and inhibitory functions for BTLA. Active programs are currently in the process of evaluating therapies designed to target these receptors in cancer. Less is understood about CD160 in human disease due to its contrasting functions in cytotoxic $CD8^+$ T cells, innate NK and $\gamma\delta$ T cells *versus* $CD4^+$ T cells. However, CD160 remains a viable pro-inflammatory target based on preliminary *in vivo* studies. Several explanations have been proposed for the tumor suppressor functions of HVEM including BTLA inhibition of Tfh cells in the context of lymphoma, activation of CD160 in cytotoxic T cells and NK and $\gamma\delta$ T cells, and direct induction of apoptosis in tumors. We additionally suggest that HVEM may control cell-intrinsic oncogenic functions through regulation of TRAF2/TRAF3 complexes, PI3K activation, or other downstream signals yet undescribed. Further understanding of how the HVEM network functions in different cancer contexts will require experimental dissection of these molecular pathways, and how lesions in this network promote oncogenesis. Appreciation of how the HVEM network functions in cancer will allow for development of individually tailored immunostimulatory or oncolytic therapies.

Funding

This work was supported by a Sanford Burnham Prebys-STRIVE Innovation Grant (to J.R.S.).

References

Abate-Daga, D., Hanada, K., Davis, J. L., Yang, J. C., Rosenberg, S. A., & Morgan, R. A. (2013). Expression profiling of TCR-engineered T cells demonstrates overexpression of multiple inhibitory receptors in persisting lymphocytes. *Blood, 122*(8), 1399–1410. https://doi.org/10.1182/blood-2013-04-495531.

Alcami, A. (2003). Viral mimicry of cytokines, chemokines and their receptors. *Nature Reviews. Immunology, 3*(1), 36–50.

Aquino-Lopez, A., Senyukov, V. V., Vlasic, Z., Kleinerman, E. S., & Lee, D. A. (2017). Interferon gamma induces changes in natural killer (NK) cell ligand expression and alters NK cell-mediated lysis of pediatric Cancer cell lines. *Frontiers in Immunology, 8*, 391. https://doi.org/10.3389/fimmu.2017.00391.

Attanasio, J., & Wherry, E. J. (2016). Costimulatory and coinhibitory receptor pathways in infectious disease. *Immunity, 44*(5), 1052–1068. https://doi.org/10.1016/j.immuni.2016.04.022.

Basso, K., & Dalla-Favera, R. (2015). Germinal centres and B cell lymphomagenesis. *Nature Reviews. Immunology, 15*(3), 172–184. https://doi.org/10.1038/nri3814.

Bechill, J., & Muller, W. J. (2014). Herpesvirus entry mediator (HVEM) attenuates signals mediated by the lymphotoxin beta receptor (LTbetaR) in human cells stimulated by the shared ligand LIGHT. *Molecular Immunology, 62*(1), 96–103. https://doi.org/10.1016/j.molimm.2014.06.013.

Bekiaris, V., Sedy, J. R., Macauley, M. G., Rhode-Kurnow, A., & Ware, C. F. (2013). The inhibitory receptor BTLA controls gammadelta T cell homeostasis and inflammatory responses. *Immunity, 39*(6), 1082–1094. https://doi.org/10.1016/j.immuni.2013.10.017.

Bertoni, F., Rossi, D., & Zucca, E. (2018). Recent advances in understanding the biology of marginal zone lymphoma. *F1000Res, 7*, 406. https://doi.org/10.12688/f1000research.13826.1.

Bienkowska, J., Allaire, N., Thai, A., Goyal, J., Plavina, T., Nirula, A., et al. (2014). Lymphotoxin-LIGHT pathway regulates the interferon signature in rheumatoid arthritis. *PLoS One*, *9*(11), e112545. https://doi.org/10.1371/journal.pone.0112545.

Bjordahl, R. L., Steidl, C., Gascoyne, R. D., & Ware, C. F. (2013). Lymphotoxin network pathways shape the tumor microenvironment. *Current Opinion in Immunology*, *25*(2), 222–229. https://doi.org/10.1016/j.coi.2013.01.001.

Bodmer, J. L., Schneider, P., & Tschopp, J. (2002). The molecular architecture of the TNF superfamily. *Trends in Biochemical Sciences*, *27*(1), 19–26.

Boice, M., Salloum, D., Mourcin, F., Sanghvi, V., Amin, R., Oricchio, E., et al. (2016). Loss of the HVEM tumor suppressor in lymphoma and restoration by modified CAR-T cells. *Cell*, *167*(2). https://doi.org/10.1016/j.cell.2016.08.032. 405–418.e13.

Bossen, C., Ingold, K., Tardivel, A., Bodmer, J. L., Gaide, O., Hertig, S., et al. (2006). Interactions of tumor necrosis factor (TNF) and TNF receptor family members in the mouse and human. *The Journal of Biological Chemistry*, *281*(20), 13964–13971.

Brunetti, G., Rizzi, R., Oranger, A., Gigante, I., Mori, G., Taurino, G., et al. (2014). LIGHT/TNFSF14 increases osteoclastogenesis and decreases osteoblastogenesis in multiple myeloma-bone disease. *Oncotarget*, *5*(24), 12950–12967. https://doi.org/10.18632/oncotarget.2633.

Cai, G., Anumanthan, A., Brown, J. A., Greenfield, E. A., Zhu, B., & Freeman, G. J. (2008). CD160 inhibits activation of human CD4+ T cells through interaction with herpesvirus entry mediator. *Nature Immunology*, *9*(2), 176–185. https://doi.org/10.1038/ni1554.

Callahan, M. K., Wolchok, J. D., Allison, J. P., & Sharma, P. (2013). Immune co-signaling to treat cancer. In *Cancer immunotherapy* (pp. 211–280): Springer.

Cao, C., Liu, S., Lou, S. F., & Liu, T. (2014). The +252A/G polymorphism in the lymphotoxin-alpha gene and the risk of non-Hodgkin lymphoma: A meta-analysis. *European Review for Medical and Pharmacological Sciences*, *18*(4), 544–552.

Carreras, J., Lopez-Guillermo, A., Fox, B. C., Colomo, L., Martinez, A., Roncador, G., et al. (2006). High numbers of tumor-infiltrating FOXP3-positive regulatory T cells are associated with improved overall survival in follicular lymphoma. *Blood*, *108*(9), 2957–2964. https://doi.org/10.1182/blood-2006-04-018218.

Carreras, J., Lopez-Guillermo, A., Roncador, G., Villamor, N., Colomo, L., Martinez, A., et al. (2009). High numbers of tumor-infiltrating programmed cell death 1-positive regulatory lymphocytes are associated with improved overall survival in follicular lymphoma. *Journal of Clinical Oncology*, *27*(9), 1470–1476. https://doi.org/10.1200/JCO.2008.18.0513.

Carter, P. J., & Lazar, G. A. (2018). Next generation antibody drugs: Pursuit of the 'high-hanging fruit'. *Nature Reviews. Drug Discovery*, *17*(3), 197–223. https://doi.org/10.1038/nrd.2017.227.

Casulo, C., Burack, W. R., & Friedberg, J. W. (2015). Transformed follicular non-Hodgkin lymphoma. *Blood*, *125*(1), 40–47. https://doi.org/10.1182/blood-2014-04-516815.

Cattoretti, G., Pasqualucci, L., Ballon, G., Tam, W., Nandula, S. V., Shen, Q., et al. (2005). Deregulated BCL6 expression recapitulates the pathogenesis of human diffuse large B cell lymphomas in mice. *Cancer Cell*, *7*(5), 445–455. https://doi.org/10.1016/j.ccr.2005.03.037.

Chabot, S., Jabrane-Ferrat, N., Bigot, K., Tabiasco, J., Provost, A., Golzio, M., et al. (2011). A novel antiangiogenic and vascular normalization therapy targeted against human CD160 receptor. *The Journal of Experimental Medicine*, *208*(5), 973–986. https://doi.org/10.1084/jem.20100810.

Chen, W., Lv, X., Liu, C., Chen, R., Liu, J., Dai, H., et al. (2018). Hematopoietic stem/progenitor cell differentiation towards myeloid lineage is modulated by LIGHT/LIGHT receptor signaling. *Journal of Cellular Physiology*, *233*(2), 1095–1103. https://doi.org/10.1002/jcp.25967.

Cheung, T. C., Humphreys, I. R., Potter, K. G., Norris, P. S., Shumway, H. M., Tran, B. R., et al. (2005). Evolutionarily divergent herpesviruses modulate T cell activation by targeting the herpesvirus entry mediator cosignaling pathway. *Proceedings of the National Academy of Sciences of the United States of America*, *102*(37), 13218–13223.

Cheung, K. J., Johnson, N. A., Affleck, J. G., Severson, T., Steidl, C., Ben-Neriah, S., et al. (2010). Acquired TNFRSF14 mutations in follicular lymphoma are associated with worse prognosis. *Cancer Research*, *70*(22), 9166–9174. https://doi.org/10.1158/0008-5472.CAN-10-2460.

Cheung, T. C., Oborne, L. M., Steinberg, M. W., Macauley, M. G., Fukuyama, S., Sanjo, H., et al. (2009). T cell intrinsic heterodimeric complexes between HVEM and BTLA determine receptivity to the surrounding microenvironment. *Journal of Immunology*, *183*(11), 7286–7296. https://doi.org/10.4049/jimmunol.0902490.

Cheung, T. C., Steinberg, M. W., Oborne, L. M., Macauley, M. G., Fukuyama, S., Sanjo, H., et al. (2009). Unconventional ligand activation of herpesvirus entry mediator signals cell survival. *Proceedings of the National Academy of Sciences of the United States of America*, *106*(15), 6244–6249. https://doi.org/10.1073/pnas.0902115106.

Chevalier, M. F., Bohner, P., Pieraerts, C., Lhermitte, B., Gourmaud, J., Nobile, A., et al. (2017). Immunoregulation of dendritic cell subsets by inhibitory receptors in urothelial cancer. *European Urology*, *71*(6), 854–857. https://doi.org/10.1016/j.eururo.2016.10.009.

Costello, R. T., Mallet, F., Barbarat, B., Schiano De Colella, J. M., Sainty, D., Sweet, R. W., et al. (2003). Stimulation of non-Hodgkin's lymphoma via HVEM: An alternate and safe way to increase Fas-induced apoptosis and improve tumor immunogenicity. *Leukemia*, *17*(12), 2500–2507.

Croft, M., Benedict, C. A., & Ware, C. F. (2013). Clinical targeting of the TNF and TNFR superfamilies. *Nature Reviews. Drug Discovery*, *12*(2), 147–168. https://doi.org/10.1038/nrd3930.

Croft, M., & Siegel, R. M. (2017). Beyond TNF: TNF superfamily cytokines as targets for the treatment of rheumatic diseases. *Nature Reviews Rheumatology*, *13*(4), 217–233. https://doi.org/10.1038/nrrheum.2017.22.

Derre, L., Rivals, J. P., Jandus, C., Pastor, S., Rimoldi, D., Romero, P., et al. (2010). BTLA mediates inhibition of human tumor-specific CD8 + T cells that can be partially reversed by vaccination. *The Journal of Clinical Investigation*, *120*(1), 157–167. https://doi.org/10.1172/JCI40070.

Ding, Y., Shen, J., Zhang, G., Chen, X., Wu, J., & Chen, W. (2015). CD40 controls CXCR5-induced recruitment of myeloid-derived suppressor cells to gastric cancer. *Oncotarget*, *6*(36), 38901–38911. https://doi.org/10.18632/oncotarget.5644.

Doering, T. A., Crawford, A., Angelosanto, J. M., Paley, M. A., Ziegler, C. G., & Wherry, E. J. (2012). Network analysis reveals centrally connected genes and pathways involved in CD8+ T cell exhaustion versus memory. *Immunity*, *37*(6), 1130–1144. https://doi.org/10.1016/j.immuni.2012.08.021.

Domae, E., Hirai, Y., Ikeo, T., Goda, S., & Shimizu, Y. (2017). Cytokine-mediated activation of human ex vivo-expanded Vgamma9Vdelta2 T cells. *Oncotarget*, *8*(28), 45928–45942. https://doi.org/10.18632/oncotarget.17498.

Dominguez-Sola, D., Kung, J., Holmes, A. B., Wells, V. A., Mo, T., Basso, K., et al. (2015). The FOXO1 transcription factor instructs the germinal center dark zone program. *Immunity*, *43*(6), 1064–1074. https://doi.org/10.1016/j.immuni.2015.10.015.

Fan, Z., Yu, P., Wang, Y., Wang, Y., Fu, M. L., Liu, W., et al. (2006). NK-cell activation by LIGHT triggers tumor-specific CD8+ T-cell immunity to reject established tumors. *Blood*, *107*(4), 1342–1351.

Fang, Y., Ye, L., Zhang, T., He, Q. Z., & Zhu, J. L. (2017). High expression of herpesvirus entry mediator (HVEM) in ovarian serous adenocarcinoma tissue. *Journal of BUON*, *22*(1), 80–86.

Farren, T. W., Giustiniani, J., Fanous, M., Liu, F., Macey, M. G., Wright, F., et al. (2015). Minimal residual disease detection with tumor-specific CD160 correlates with event-free survival in chronic lymphocytic leukemia. *Blood Cancer Journal*, *5*, e273. https://doi.org/10.1038/bcj.2014.92.

Farren, T. W., Giustiniani, J., Liu, F. T., Tsitsikas, D. A., Macey, M. G., Cavenagh, J. D., et al. (2011). Differential and tumor-specific expression of CD160 in B-cell malignancies. *Blood*, *118*(8), 2174–2183. https://doi.org/10.1182/blood-2011-02-334326.

Feng, X. Y., Wen, X. Z., Tan, X. J., Hou, J. H., Ding, Y., Wang, K. F., et al. (2015). Ectopic expression of B and T lymphocyte attenuator in gastric cancer: A potential independent prognostic factor in patients with gastric cancer. *Molecular Medicine Reports*, *11*(1), 658–664. https://doi.org/10.3892/mmr.2014.2699.

Fhu, C. W., Graham, A. M., Yap, C. T., Al-Salam, S., Castella, A., Chong, S. M., et al. (2014). Reed-Sternberg cell-derived lymphotoxin-alpha activates endothelial cells to enhance T-cell recruitment in classical Hodgkin lymphoma. *Blood*, *124*(19), 2973–2982. https://doi.org/10.1182/blood-2014-05-576140.

Flynn, R., Hutchinson, T., Murphy, K. M., Ware, C. F., Croft, M., & Salek-Ardakani, S. (2013). CD8 T cell memory to a viral pathogen requires trans cosignaling between HVEM and BTLA. *PLoS One*, *8*(10), e77991. https://doi.org/10.1371/journal.pone.0077991.

Fons, P., Chabot, S., Cartwright, J. E., Lenfant, F., L'Faqihi, F., Giustiniani, J., et al. (2006). Soluble HLA-G1 inhibits angiogenesis through an apoptotic pathway and by direct binding to CD160 receptor expressed by endothelial cells. *Blood*, *108*(8), 2608–2615. https://doi.org/10.1182/blood-2005-12-019919.

Forbes, S. A., Beare, D., Boutselakis, H., Bamford, S., Bindal, N., Tate, J., et al. (2017). COSMIC: Somatic cancer genetics at high-resolution. *Nucleic Acids Research*, *45*(D1), D777–D783. https://doi.org/10.1093/nar/gkw1121.

Forget, M. A., Haymaker, C., Hess, K. R., Meng, Y. J., Creasy, C., Karpinets, T., et al. (2018). Prospective analysis of adoptive TIL therapy in patients with metastatic melanoma: Response, impact of anti-CTLA4, and biomarkers to predict clinical outcome. *Clinical Cancer Research*, *24*, 4416–4428. https://doi.org/10.1158/1078-0432.CCR-17-3649.

Fourcade, J., Sun, Z., Pagliano, O., Guillaume, P., Luescher, I. F., Sander, C., et al. (2012). CD8(+) T cells specific for tumor antigens can be rendered dysfunctional by the tumor microenvironment through upregulation of the inhibitory receptors BTLA and PD-1. *Cancer Research*, *72*(4), 887–896. https://doi.org/10.1158/0008-5472.CAN-11-2637.

Gango, A., Batai, B., Varga, M., Kapczar, D., Papp, G., Marschalko, M., et al. (2018). Concomitant 1p36 deletion and TNFRSF14 mutations in primary cutaneous follicle center lymphoma frequently expressing high levels of EZH2 protein. *Virchows Archiv*, *473*, 453–462. https://doi.org/10.1007/s00428-018-2384-3.

Gao, J., Aksoy, B. A., Dogrusoz, U., Dresdner, G., Gross, B., Sumer, S. O., et al. (2013). Integrative analysis of complex cancer genomics and clinical profiles using the cBioPortal. *Science Signaling*, *6*(269), pl1. https://doi.org/10.1126/scisignal.2004088.

Gavrieli, M., & Murphy, K. M. (2006). Association of Grb-2 and PI3K p85 with phosphotyrosile peptides derived from BTLA. *Biochemical and Biophysical Research Communications*, *345*(4), 1440–1445.

Ge, J., Zhu, L., Zhou, J., Li, G., Li, Y., Li, S., et al. (2015). Association between co-inhibitory molecule gene tagging single nucleotide polymorphisms and the risk of colorectal cancer in Chinese. *Journal of Cancer Research and Clinical Oncology*, *141*(9), 1533–1544. https://doi.org/10.1007/s00432-015-1915-4.

Gertner-Dardenne, J., Fauriat, C., Orlanducci, F., Thibult, M. L., Pastor, S., Fitzgibbon, J., et al. (2013). The co-receptor BTLA negatively regulates human Vgamma9Vdelta2 T-cell proliferation: A potential way of immune escape for lymphoma cells. *Blood*, *122*(6), 922–931. https://doi.org/10.1182/blood-2012-11-464685.

Ghazavi, F., Clappier, E., Lammens, T., Suciu, S., Caye, A., Zegrari, S., et al. (2015). CD200/BTLA deletions in pediatric precursor B-cell acute lymphoblastic leukemia treated according to the EORTC-CLG 58951 protocol. *Haematologica, 100*(10), 1311–1319. https://doi.org/10.3324/haematol.2015.126953.

Gong, L. L., Han, F. F., Lv, Y. L., Liu, H., Wan, Z. R., Zhang, W., et al. (2017). TNF-alpha and LT-alpha polymorphisms and the risk of leukemia: A meta-analysis. *Tumori, 103*(1), 53–59. https://doi.org/10.5301/tj.5000549.

Green, M. R., Gentles, A. J., Nair, R. V., Irish, J. M., Kihira, S., Liu, C. L., et al. (2013). Hierarchy in somatic mutations arising during genomic evolution and progression of follicular lymphoma. *Blood, 121*(9), 1604–1611. https://doi.org/10.1182/blood-2012-09-457283.

Han, L., Wang, W., Lu, J., Kong, F., Ma, G., Zhu, Y., et al. (2014). AAV-sBTLA facilitates HSP70 vaccine-triggered prophylactic antitumor immunity against a murine melanoma pulmonary metastasis model in vivo. *Cancer Letters, 354*(2), 398–406. https://doi.org/10.1016/j.canlet.2014.08.006.

Haymaker, C. L., Wu, R. C., Ritthipichai, K., Bernatchez, C., Forget, M. A., Chen, J. Q., et al. (2015). BTLA marks a less-differentiated tumor-infiltrating lymphocyte subset in melanoma with enhanced survival properties. *Oncoimmunology, 4*(8), e1014246. https://doi.org/10.1080/2162402X.2015.1014246.

He, B., Jabouille, A., Steri, V., Johansson-Percival, A., Michael, I. P., Kotamraju, V. R., et al. (2018). Vascular targeting of LIGHT normalizes blood vessels in primary brain cancer and induces intratumoural high endothelial venules. *The Journal of Pathology, 245*(2), 209–221. https://doi.org/10.1002/path.5080.

Heo, S. K., Noh, E. K., Gwon, G. D., Kim, J. Y., Jo, J. C., Choi, Y., et al. (2016). LIGHT (TNFSF14) increases the survival and proliferation of human bone marrow-derived mesenchymal stem cells. *PLoS One, 11*(11), e0166589. https://doi.org/10.1371/journal.pone.0166589.

Herro, R., Antunes, R. D. S., Aguilera, A. R., Tamada, K., & Croft, M. (2015). The tumor necrosis factor superfamily molecule LIGHT promotes keratinocyte activity and skin fibrosis. *The Journal of Investigative Dermatology, 135*(8), 2109–2118. https://doi.org/10.1038/jid.2015.110.

Herro, R., & Croft, M. (2016). The control of tissue fibrosis by the inflammatory molecule LIGHT (TNF superfamily member 14). *Pharmacological Research, 104*, 151–155. https://doi.org/10.1016/j.phrs.2015.12.018.

Herro, R., Da Silva Antunes, R., Aguilera, A. R., Tamada, K., & Croft, M. (2015). Tumor necrosis factor superfamily 14 (LIGHT) controls thymic stromal lymphopoietin to drive pulmonary fibrosis. *The Journal of Allergy and Clinical Immunology, 136*(3), 757–768. https://doi.org/10.1016/j.jaci.2014.12.1936.

Herro, R., Shui, J. W., Zahner, S., Sidler, D., Kawakami, Y., Kawakami, T., et al. (2018). LIGHT-HVEM signaling in keratinocytes controls development of dermatitis. *The Journal of Experimental Medicine, 215*(2), 415–422. https://doi.org/10.1084/jem.20170536.

Hokuto, D., Sho, M., Yamato, I., Yasuda, S., Obara, S., Nomi, T., et al. (2015). Clinical impact of herpesvirus entry mediator expression in human hepatocellular carcinoma. *European Journal of Cancer, 51*(2), 157–165. https://doi.org/10.1016/j.ejca.2014.11.004.

Holmes, T. D., Wilson, E. B., Black, E. V., Benest, A. V., Vaz, C., Tan, B., et al. (2014). Licensed human natural killer cells aid dendritic cell maturation via TNFSF14/LIGHT. *Proceedings of the National Academy of Sciences of the United States of America, 111*(52), E5688–E5696. https://doi.org/10.1073/pnas.1411072112.

Hu, X., Zimmerman, M. A., Bardhan, K., Yang, D., Waller, J. L., Liles, G. B., et al. (2013). Lymphotoxin beta receptor mediates caspase-dependent tumor cell apoptosis in vitro and tumor suppression in vivo despite induction of NF-kappaB activation. *Carcinogenesis, 34*(5), 1105–1114. https://doi.org/10.1093/carcin/bgt014.

Huang, Y., Yu, X., Wang, L., Zhou, S., Sun, J., Feng, N., et al. (2013). Four genetic polymorphisms of lymphotoxin-alpha gene and cancer risk: A systematic review and meta-analysis. *PLoS One, 8*(12), e82519. https://doi.org/10.1371/journal.pone.0082519.

Inoue, T., Sho, M., Yasuda, S., Nishiwada, S., Nakamura, S., Ueda, T., et al. (2015). HVEM expression contributes to tumor progression and prognosis in human colorectal cancer. *Anticancer Research, 35*(3), 1361–1367.

Johansson-Percival, A., He, B., Li, Z. J., Kjellen, A., Russell, K., Li, J., et al. (2017). De novo induction of intratumoral lymphoid structures and vessel normalization enhances immunotherapy in resistant tumors. *Nature Immunology, 18*(11), 1207–1217. https://doi.org/10.1038/ni.3836.

Johansson-Percival, A., Li, Z. J., Lakhiani, D. D., He, B., Wang, X., Hamzah, J., et al. (2015). Intratumoral LIGHT restores pericyte contractile properties and vessel integrity. *Cell Reports, 13*(12), 2687–2698. https://doi.org/10.1016/j.celrep.2015.12.004.

Kabrani, E., Chu, V. T., Tasouri, E., Sommermann, T., Bassler, K., Ulas, T., et al. (2018). Nuclear FOXO1 promotes lymphomagenesis in germinal center B cells. *Blood, 132*, 2670–2683. https://doi.org/10.1182/blood-2018-06-856203.

Karabon, L., Partyka, A., Jasek, M., Lech-Maranda, E., Grzybowska-Izydorczyk, O., Bojarska-Junak, A., et al. (2016). Intragenic variations in BTLA gene influence mRNA expression of BTLA gene in chronic lymphocytic leukemia patients and confer susceptibility to chronic lymphocytic leukemia. *Archivum Immunologiae et Therapiae Experimentalis (Warsz), 64*(Suppl. 1), 137–145. https://doi.org/10.1007/s00005-016-0430-x.

Kennedy, W. P., Simon, J. A., Offutt, C., Horn, P., Herman, A., Townsend, M. J., et al. (2014). Efficacy and safety of pateclizumab (anti-lymphotoxin-alpha) compared to adalimumab in rheumatoid arthritis: A head-to-head phase 2 randomized controlled study (the ALTARA study). *Arthritis Research & Therapy, 16*(5), 467. https://doi.org/10.1186/s13075-014-0467-3.

Klingemann, H. G. (2013). Cellular therapy of cancer with natural killer cells-where do we stand? *Cytotherapy, 15*(10), 1185–1194. https://doi.org/10.1016/j.jcyt.2013.03.011.

Kobayashi, Y., Iwata, A., Suzuki, K., Suto, A., Kawashima, S., Saito, Y., et al. (2013). B and T lymphocyte attenuator inhibits LPS-induced endotoxic shock by suppressing toll-like receptor 4 signaling in innate immune cells. *Proceedings of the National Academy of Sciences of the United States of America, 110*(13), 5121–5126. https://doi.org/10.1073/pnas.1222093110.

Kojima, R., Kajikawa, M., Shiroishi, M., Kuroki, K., & Maenaka, K. (2011). Molecular basis for herpesvirus entry mediator recognition by the human immune inhibitory receptor CD160 and its relationship to the cosignaling molecules BTLA and LIGHT. *Journal of Molecular Biology, 413*(4), 762–772. https://doi.org/10.1016/j.jmb.2011.09.018.

Kotsiou, E., Okosun, J., Besley, C., Iqbal, S., Matthews, J., Fitzgibbon, J., et al. (2016). TNFRSF14 aberrations in follicular lymphoma increase clinically significant allogeneic T-cell responses. *Blood, 128*(1), 72–81. https://doi.org/10.1182/blood-2015-10-679191.

Lan, X., Li, S., Gao, H., Nanding, A., Quan, L., Yang, C., et al. (2017). Increased BTLA and HVEM in gastric cancer are associated with progression and poor prognosis. *Onco Targets and Therapy, 10*, 919–926. https://doi.org/10.2147/OTT.S128825.

Landau, D. A., Tausch, E., Taylor-Weiner, A. N., Stewart, C., Reiter, J. G., Bahlo, J., et al. (2015). Mutations driving CLL and their evolution in progression and relapse. *Nature, 526*(7574), 525–530. https://doi.org/10.1038/nature15395.

Lauenborg, B., Christensen, L., Ralfkiaer, U., Kopp, K. L., Jonson, L., Dabelsteen, S., et al. (2015). Malignant T cells express lymphotoxin alpha and drive endothelial activation in cutaneous T cell lymphoma. *Oncotarget, 6*(17), 15235–15249. https://doi.org/10.18632/oncotarget.3837.

Launay, E., Pangault, C., Bertrand, P., Jardin, F., Lamy, T., Tilly, H., et al. (2012). High rate of TNFRSF14 gene alterations related to 1p36 region in de novo follicular lymphoma and impact on prognosis. *Leukemia, 26*(3), 559–562. https://doi.org/10.1038/leu.2011.266.

Lee, E. H., Kim, E. M., Ji, K. Y., Park, A. R., Choi, H. R., Lee, H. Y., et al. (2017). Axl acts as a tumor suppressor by regulating LIGHT expression in T lymphoma. *Oncotarget, 8*(13), 20645–20655. https://doi.org/10.18632/oncotarget.15830.

Legat, A., Speiser, D. E., Pircher, H., Zehn, D., & Fuertes Marraco, S. A. (2013). Inhibitory receptor expression depends more dominantly on differentiation and activation than "exhaustion" of human CD8 T cells. *Frontiers in Immunology, 4*, 455. https://doi.org/10.3389/fimmu.2013.00455.

Lesesve, J. F., Tardy, S., Frotscher, B., Latger-Cannard, V., Feugier, P., & De Carvalho Bittencourt, M. (2015). Combination of CD160 and CD200 as a useful tool for differential diagnosis between chronic lymphocytic leukemia and other mature B-cell neoplasms. *International Journal of Laboratory Hematology, 37*(4), 486–494. https://doi.org/10.1111/ijlh.12315.

Li, D., Fu, Z., Chen, S., Yuan, W., Liu, Y., Li, L., et al. (2013). HVEM gene polymorphisms are associated with sporadic breast cancer in Chinese women. *PLoS One, 8*(8), e71040. https://doi.org/10.1371/journal.pone.0071040.

Li, Q., Johnston, N., Zheng, X., Wang, H., Zhang, X., Gao, D., et al. (2016). miR-28 modulates exhaustive differentiation of T cells through silencing programmed cell death-1 and regulating cytokine secretion. *Oncotarget, 7*(33), 53735–53750. https://doi.org/10.18632/oncotarget.10731.

Lichtenegger, F. S., Kondla, I., Krempasky, M., Weber, A. L., Herold, T., Krupka, C., et al. (2015). RNA and protein expression of herpesvirus entry mediator (HVEM) is associated with molecular markers, immunity-related pathways and relapse-free survival of patients with AML. *Cancer Immunology, Immunotherapy, 64*(12), 1505–1515. https://doi.org/10.1007/s00262-015-1755-8.

Liu, F. T., Giustiniani, J., Farren, T., Jia, L., Bensussan, A., Gribben, J. G., et al. (2010). CD160 signaling mediates PI3K-dependent survival and growth signals in chronic lymphocytic leukemia. *Blood, 115*(15), 3079–3088. https://doi.org/10.1182/blood-2009-08-239483.

Liu, Y. F., Wang, B. Y., Zhang, W. N., Huang, J. Y., Li, B. S., Zhang, M., et al. (2016). Genomic profiling of adult and pediatric B-cell acute lymphoblastic leukemia. *eBioMedicine, 8*, 173–183. https://doi.org/10.1016/j.ebiom.2016.04.038.

Liu, W., Zhan, C., Cheng, H., Kumar, P. R., Bonanno, J. B., Nathenson, S. G., et al. (2014). Mechanistic basis for functional promiscuity in the TNF and TNF receptor superfamilies: Structure of the LIGHT:DcR3 assembly. *Structure, 22*(9), 1252–1262. https://doi.org/10.1016/j.str.2014.06.013.

Lohr, J. G., Stojanov, P., Lawrence, M. S., Auclair, D., Chapuy, B., Sougnez, C., et al. (2012). Discovery and prioritization of somatic mutations in diffuse large B-cell lymphoma (DLBCL) by whole-exome sequencing. *Proceedings of the National Academy of Sciences of the United States of America, 109*(10), 3879–3884. https://doi.org/10.1073/pnas.1121343109.

Long, M., Beckwith, K., Do, P., Mundy, B. L., Gordon, A., Lehman, A. M., et al. (2017). Ibrutinib treatment improves T cell number and function in CLL patients. *The Journal of Clinical Investigation, 127*(8), 3052–3064. https://doi.org/10.1172/JCI89756.

Louissaint, A., Jr., Schafernak, K. T., Geyer, J. T., Kovach, A. E., Ghandi, M., Gratzinger, D., et al. (2016). Pediatric-type nodal follicular lymphoma: A biologically distinct lymphoma with frequent MAPK pathway mutations. *Blood, 128*(8), 1093–1100. https://doi.org/10.1182/blood-2015-12-682591.

Maker, A. V., Ito, H., Mo, Q., Weisenberg, E., Qin, L. X., Turcotte, S., et al. (2015). Genetic evidence that intratumoral T-cell proliferation and activation are associated with recurrence and survival in patients with resected colorectal liver metastases. *Cancer Immunology Research*, *3*(4), 380–388. https://doi.org/10.1158/2326-6066.CIR-14-0212.

Martin-Guerrero, I., Salaverria, I., Burkhardt, B., Szczepanowski, M., Baudis, M., Bens, S., et al. (2013). Recurrent loss of heterozygosity in 1p36 associated with TNFRSF14 mutations in IRF4 translocation negative pediatric follicular lymphomas. *Haematologica*, *98*(8), 1237–1241. https://doi.org/10.3324/haematol.2012.073916.

McDonald, B. D., Jabri, B., & Bendelac, A. (2018). Diverse developmental pathways of intestinal intraepithelial lymphocytes. *Nature Reviews. Immunology*, *18*(8), 514–525. https://doi.org/10.1038/s41577-018-0013-7.

Migita, K., Sho, M., Shimada, K., Yasuda, S., Yamato, I., Takayama, T., et al. (2014). Significant involvement of herpesvirus entry mediator in human esophageal squamous cell carcinoma. *Cancer*, *120*(6), 808–817. https://doi.org/10.1002/cncr.28491.

Miletic, A. V., Anzelon-Mills, A. N., Mills, D. M., Omori, S. A., Pedersen, I. M., Shin, D. M., et al. (2010). Coordinate suppression of B cell lymphoma by PTEN and SHIP phosphatases. *The Journal of Experimental Medicine*, *207*(11), 2407–2420. https://doi.org/10.1084/jem.20091962.

Miyagaki, T., Sugaya, M., Suga, H., Ohmatsu, H., Fujita, H., Asano, Y., et al. (2013). Serum-soluble herpes virus entry mediator levels reflect disease severity and Th2 environment in cutaneous T-cell lymphoma. *Acta Dermato-Venereologica*, *93*(4), 465–467. https://doi.org/10.2340/00015555-1523.

Moffitt, A. B., & Dave, S. S. (2017). Clinical applications of the genomic landscape of aggressive non-Hodgkin lymphoma. *Journal of Clinical Oncology*, *35*(9), 955–962. https://doi.org/10.1200/JCO.2016.71.7603.

Moody, S., Thompson, J. S., Chuang, S. S., Liu, H., Raderer, M., Vassiliou, G., et al. (2018). Novel GPR34 and CCR6 mutation and distinct genetic profiles in MALT lymphomas of different sites. *Haematologica*, *103*(8), 1329–1336. https://doi.org/10.3324/haematol.2018.191601.

Morin, R. D., Mendez-Lago, M., Mungall, A. J., Goya, R., Mungall, K. L., Corbett, R. D., et al. (2011). Frequent mutation of histone-modifying genes in non-Hodgkin lymphoma. *Nature*, *476*(7360), 298–303. https://doi.org/10.1038/nature10351.

Morishige, T., Yoshioka, Y., Narimatsu, S., Ikemizu, S., Tsunoda, S., Tsutsumi, Y., et al. (2013). Mutants of lymphotoxin-alpha with augmented cytotoxic activity via TNFR1 for use in cancer therapy. *Cytokine*, *61*(2), 578–584. https://doi.org/10.1016/j.cyto.2012.11.005.

Mullighan, C. G., Goorha, S., Radtke, I., Miller, C. B., Coustan-Smith, E., Dalton, J. D., et al. (2007). Genome-wide analysis of genetic alterations in acute lymphoblastic leukaemia. *Nature*, *446*(7137), 758–764. https://doi.org/10.1038/nature05690.

Murphy, T. L., & Murphy, K. M. (2010). Slow down and survive: Enigmatic immunoregulation by BTLA and HVEM. *Annual Review of Immunology*, *28*, 389–411. https://doi.org/10.1146/annurev-immunol-030409-101202.

Murphy, K. M., Nelson, C. A., & Sedy, J. R. (2006). Balancing co-stimulation and inhibition with BTLA and HVEM. *Nature Reviews Immunology*, *6*(9), 671–681.

Muschen, M. (2018). Autoimmunity checkpoints as therapeutic targets in B cell malignancies. *Nature Reviews. Cancer*, *18*(2), 103–116. https://doi.org/10.1038/nrc.2017.111.

Oguro, S., Ino, Y., Shimada, K., Hatanaka, Y., Matsuno, Y., Esaki, M., et al. (2015). Clinical significance of tumor-infiltrating immune cells focusing on BTLA and Cbl-b in patients with gallbladder cancer. *Cancer Science*, *106*(12), 1750–1760. https://doi.org/10.1111/cas.12825.

Okosun, J., Wolfson, R. L., Wang, J., Araf, S., Wilkins, L., Castellano, B. M., et al. (2016). Recurrent mTORC1-activating RRAGC mutations in follicular lymphoma. *Nature Genetics*, *48*(2), 183–188. https://doi.org/10.1038/ng.3473.

Ortega-Molina, A., Boss, I. W., Canela, A., Pan, H., Jiang, Y., Zhao, C., et al. (2015). The histone lysine methyltransferase KMT2D sustains a gene expression program that represses B cell lymphoma development. *Nature Medicine, 21*(10), 1199–1208. https://doi.org/10.1038/nm.3943.

Otterdal, K., Haukeland, J. W., Yndestad, A., Dahl, T. B., Holm, S., Segers, F. M., et al. (2015). Increased serum levels of LIGHT/TNFSF14 in nonalcoholic fatty liver disease: Possible role in hepatic inflammation. *Clinical and Translational Gastroenterology, 6*, e95. https://doi.org/10.1038/ctg.2015.23.

Pangault, C., Ame-Thomas, P., Ruminy, P., Rossille, D., Caron, G., Baia, M., et al. (2010). Follicular lymphoma cell niche: Identification of a preeminent IL-4-dependent T(FH)-B cell axis. *Leukemia, 24*(12), 2080–2089. https://doi.org/10.1038/leu.2010.223.

Pardoll, D. M. (2012). The blockade of immune checkpoints in cancer immunotherapy. *Nature Reviews. Cancer, 12*(4), 252–264. https://doi.org/10.1038/nrc3239.

Partyka, A., Tupikowski, K., Kolodziej, A., Zdrojowy, R., Halon, A., Malkiewicz, B., et al. (2016). Association of 3′ nearby gene BTLA polymorphisms with the risk of renal cell carcinoma in the Polish population. *Urologic Oncology, 34*(9). https://doi.org/10.1016/j.urolonc.2016.04.010. 419.e13-9.

Pasero, C., Barbarat, B., Just-Landi, S., Bernard, A., Aurran-Schleinitz, T., Rey, J., et al. (2009). A role for HVEM, but not lymphotoxin-beta receptor, in LIGHT-induced tumor cell death and chemokine production. *European Journal of Immunology, 39*(9), 2502–2514. https://doi.org/10.1002/eji.200939069.

Pasero, C., & Olive, D. (2013). Interfering with coinhibitory molecules: BTLA/HVEM as new targets to enhance anti-tumor immunity. *Immunology Letters, 151*(1-2), 71–75. https://doi.org/10.1016/j.imlet.2013.01.008.

Pasqualucci, L., & Dalla-Favera, R. (2014). SnapShot: Diffuse large B cell lymphoma. *Cancer Cell, 25*(1). https://doi.org/10.1016/j.ccr.2013.12.012. 132-132.e1.

Pasqualucci, L., Khiabanian, H., Fangazio, M., Vasishtha, M., Messina, M., Holmes, A. B., et al. (2014). Genetics of follicular lymphoma transformation. *Cell Reports, 6*(1), 130–140. https://doi.org/10.1016/j.celrep.2013.12.027.

Pasqualucci, L., Trifonov, V., Fabbri, G., Ma, J., Rossi, D., Chiarenza, A., et al. (2011). Analysis of the coding genome of diffuse large B-cell lymphoma. *Nature Genetics, 43*(9), 830–837. https://doi.org/10.1038/ng.892.

Pastore, A., Jurinovic, V., Kridel, R., Hoster, E., Staiger, A. M., Szczepanowski, M., et al. (2015). Integration of gene mutations in risk prognostication for patients receiving first-line immunochemotherapy for follicular lymphoma: A retrospective analysis of a prospective clinical trial and validation in a population-based registry. *The Lancet Oncology, 16*(9), 1111–1122. https://doi.org/10.1016/S1470-2045(15)00169-2.

Pfeifer, M., Grau, M., Lenze, D., Wenzel, S. S., Wolf, A., Wollert-Wulf, B., et al. (2013). PTEN loss defines a PI3K/AKT pathway-dependent germinal center subtype of diffuse large B-cell lymphoma. *Proceedings of the National Academy of Sciences of the United States of America, 110*(30), 12420–12425. https://doi.org/10.1073/pnas.1305656110.

Puente, X. S., Bea, S., Valdes-Mas, R., Villamor, N., Gutierrez-Abril, J., Martin-Subero, J. I., et al. (2015). Non-coding recurrent mutations in chronic lymphocytic leukaemia. *Nature, 526*(7574), 519–524. https://doi.org/10.1038/nature14666.

Qiao, G., Qin, J., Kunda, N., Calata, J. F., Mahmud, D. L., Gann, P., et al. (2017). LIGHT elevation enhances immune eradication of colon cancer metastases. *Cancer Research, 77*(8), 1880–1891. https://doi.org/10.1158/0008-5472.CAN-16-1655.

Quan, L., Lan, X., Meng, Y., Guo, X., Guo, Y., Zhao, L., et al. (2018). BTLA marks a less cytotoxic T-cell subset in diffuse large B-cell lymphoma with high expression of checkpoints. *Experimental Hematology, 60*, https://doi.org/10.1016/j.exphem.2018.01.003. 47–56.e1.

Ramezani-Rad, P., & Rickert, R. C. (2017). Murine models of germinal center derived-lymphomas. *Current Opinion in Immunology, 45*, 31–36. https://doi.org/10.1016/j.coi.2016.12.002.

Reddy, A., Zhang, J., Davis, N. S., Moffitt, A. B., Love, C. L., Waldrop, A., et al. (2017). Genetic and functional drivers of diffuse large B cell lymphoma. *Cell, 171*(2), 481–494 e15. https://doi.org/10.1016/j.cell.2017.09.027.

Riches, J. C., Davies, J. K., McClanahan, F., Fatah, R., Iqbal, S., Agrawal, S., et al. (2013). T cells from CLL patients exhibit features of T-cell exhaustion but retain capacity for cytokine production. *Blood, 121*(9), 1612–1621. https://doi.org/10.1182/blood-2012-09-457531.

Rickert, R. C. (2013). New insights into pre-BCR and BCR signalling with relevance to B cell malignancies. *Nature Reviews. Immunology, 13*(8), 578–591. https://doi.org/10.1038/nri3487.

Ritthipichai, K., Haymaker, C. L., Martinez, M., Aschenbrenner, A., Yi, X., Zhang, M., et al. (2017). Multifaceted role of BTLA in the control of CD8(+) T-cell fate after antigen encounter. *Clinical Cancer Research, 23*(20), 6151–6164. https://doi.org/10.1158/1078-0432.CCR-16-1217.

Rodriguez-Vicente, A. E., Bikos, V., Hernandez-Sanchez, M., Malcikova, J., Hernandez-Rivas, J. M., & Pospisilova, S. (2017). Next-generation sequencing in chronic lymphocytic leukemia: Recent findings and new horizons. *Oncotarget, 8*(41), 71234–71248. https://doi.org/10.18632/oncotarget.19525.

Rooney, I. A., Butrovich, K. D., Glass, A. A., Borboroglu, S., Benedict, C. A., Whitbeck, J. C., et al. (2000). The lymphotoxin-beta receptor is necessary and sufficient for LIGHT-mediated apoptosis of tumor cells. *The Journal of Biological Chemistry, 275*, 14307–14315.

Sako, N., Schiavon, V., Bounfour, T., Dessirier, V., Ortonne, N., Olive, D., et al. (2014). Membrane expression of NK receptors CD160 and CD158k contributes to delineate a unique CD4+ T-lymphocyte subset in normal and mycosis fungoides skin. *Cytometry. Part A, 85*(10), 869–882. https://doi.org/10.1002/cyto.a.22512.

Salipante, S. J., Adey, A., Thomas, A., Lee, C., Liu, Y. J., Kumar, A., et al. (2016). Recurrent somatic loss of TNFRSF14 in classical Hodgkin lymphoma. *Genes, Chromosomes & Cancer, 55*(3), 278–287. https://doi.org/10.1002/gcc.22331.

Sander, S., Chu, V. T., Yasuda, T., Franklin, A., Graf, R., Calado, D. P., et al. (2015). PI3 kinase and FOXO1 transcription factor activity differentially control B cells in the germinal center light and dark zones. *Immunity, 43*(6), 1075–1086. https://doi.org/10.1016/j.immuni.2015.10.021.

Sanjo, H., Zajonc, D. M., Braden, R., Norris, P. S., & Ware, C. F. (2010). Allosteric regulation of the ubiquitin: NIK and ubiquitin: TRAF3 E3 ligases by the lymphotoxin-beta receptor. *The Journal of Biological Chemistry, 285*(22), 17148–17155. https://doi.org/10.1074/jbc.M110.105874.

Sarhan, D., Palma, M., Mao, Y., Adamson, L., Kiessling, R., Mellstedt, H., et al. (2015). Dendritic cell regulation of NK-cell responses involves lymphotoxin-alpha, IL-12, and TGF-beta. *European Journal of Immunology, 45*(6), 1783–1793. https://doi.org/10.1002/eji.201444885.

Schmidt, J., Gong, S., Marafioti, T., Mankel, B., Gonzalez-Farre, B., Balague, O., et al. (2016). Genome-wide analysis of pediatric-type follicular lymphoma reveals low genetic complexity and recurrent alterations of TNFRSF14 gene. *Blood, 128*(8), 1101–1111. https://doi.org/10.1182/blood-2016-03-703819.

Schmidt, J., Ramis-Zaldivar, J. E., Nadeu, F., Gonzalez-Farre, B., Navarro, A., Egan, C., et al. (2017). Mutations of MAP2K1 are frequent in pediatric-type follicular lymphoma and result in ERK pathway activation. *Blood, 130*(3), 323–327. https://doi.org/10.1182/blood-2017-03-776278.

Schmitz, R., Wright, G. W., Huang, D. W., Johnson, C. A., Phelan, J. D., Wang, J. Q., et al. (2018). Genetics and pathogenesis of diffuse large B-cell lymphoma. *The New England Journal of Medicine*, *378*(15), 1396–1407. https://doi.org/10.1056/NEJMoa1801445.

Schnorfeil, F. M., Lichtenegger, F. S., Emmerig, K., Schlueter, M., Neitz, J. S., Draenert, R., et al. (2015). T cells are functionally not impaired in AML: Increased PD-1 expression is only seen at time of relapse and correlates with a shift towards the memory T cell compartment. *Journal of Hematology & Oncology*, *8*, 93. https://doi.org/10.1186/s13045-015-0189-2.

Schrama, D., thor Straten, P., Fischer, W. H., McLellan, A. D., Brocker, E. B., Reisfeld, R. A., et al. (2001). Targeting of lymphotoxin-alpha to the tumor elicits an efficient immune response associated with induction of peripheral lymphoid-like tissue. *Immunity*, *14*(2), 111–121.

Scott, D. W., & Gascoyne, R. D. (2014). The tumour microenvironment in B cell lymphomas. *Nature Reviews. Cancer*, *14*(8), 517–534. https://doi.org/10.1038/nrc3774.

Sedy, J. R., Balmert, M. O., Ware, B. C., Smith, W., Nemcovicova, I., Norris, P. S., et al. (2017). A herpesvirus entry mediator mutein with selective agonist action for the inhibitory receptor B and T lymphocyte attenuator. *Journal of Biological Chemistry*, *292*, 21060–21070. https://doi.org/10.1074/jbc.M117.813295.

Sedy, J., Bekiaris, V., & Ware, C. F. (2014). Tumor necrosis factor superfamily in innate immunity and inflammation. *Cold Spring Harbor Perspectives in Biology*, *7*(4), a016279. https://doi.org/10.1101/cshperspect.a016279.

Sedy, J. R., Bjordahl, R. L., Bekiaris, V., Macauley, M. G., Ware, B. C., Norris, P. S., et al. (2013). CD160 activation by herpesvirus entry mediator augments inflammatory cytokine production and cytolytic function by NK cells. *Journal of Immunology*, *191*(2), 828–836. https://doi.org/10.4049/jimmunol.1300894.

Sekar, D., Govene, L., Del Rio, M. L., Sirait-Fischer, E., Fink, A. F., Brune, B., et al. (2018). Downregulation of BTLA on NKT cells promotes tumor immune control in a mouse model of mammary carcinoma. *International Journal of Molecular Sciences*, *19*(3). https://doi.org/10.3390/ijms19030752.

Senft, D., Qi, J., & Ronai, Z. A. (2018). Ubiquitin ligases in oncogenic transformation and cancer therapy. *Nature Reviews. Cancer*, *18*(2), 69–88. https://doi.org/10.1038/nrc.2017.105.

Shaffer, A. L., 3rd, Young, R. M., & Staudt, L. M. (2012). Pathogenesis of human B cell lymphomas. *Annual Review of Immunology*, *30*, 565–610. https://doi.org/10.1146/annurev-immunol-020711-075027.

Shui, J. W., Larange, A., Kim, G., Vela, J. L., Zahner, S., Cheroutre, H., et al. (2012). HVEM signalling at mucosal barriers provides host defence against pathogenic bacteria. *Nature*, *488*(7410), 222–225. https://doi.org/10.1038/nature11242.

Sideras, K., Biermann, K., Verheij, J., Takkenberg, B. R., Mancham, S., Hansen, B. E., et al. (2017). PD-L1, Galectin-9 and CD8(+) tumor-infiltrating lymphocytes are associated with survival in hepatocellular carcinoma. *Oncoimmunology*, *6*(2), e1273309. https://doi.org/10.1080/2162402X.2016.1273309.

Sideras, K., Biermann, K., Yap, K., Mancham, S., Boor, P. P. C., Hansen, B. E., et al. (2017). Tumor cell expression of immune inhibitory molecules and tumor-infiltrating lymphocyte count predict cancer-specific survival in pancreatic and ampullary cancer. *International Journal of Cancer*, *141*(3), 572–582. https://doi.org/10.1002/ijc.30760.

So, T., & Croft, M. (2013). Regulation of PI-3-kinase and Akt signaling in T lymphocytes and other cells by TNFR family molecules. *Frontiers in Immunology*, *4*, 139. https://doi.org/10.3389/fimmu.2013.00139.

Soroosh, P., Doherty, T. A., So, T., Mehta, A. K., Khorram, N., Norris, P. S., et al. (2011). Herpesvirus entry mediator (TNFRSF14) regulates the persistence of T helper memory cell populations. *The Journal of Experimental Medicine*, *208*(4), 797–809. https://doi.org/10.1084/jem.20101562.

Spina, V., Khiabanian, H., Messina, M., Monti, S., Cascione, L., Bruscaggin, A., et al. (2016). The genetics of nodal marginal zone lymphoma. *Blood*, *128*(10), 1362–1373. https://doi.org/10.1182/blood-2016-02-696757.

Spodzieja, M., Lach, S., Iwaszkiewicz, J., Cesson, V., Kalejta, K., Olive, D., et al. (2017). Design of short peptides to block BTLA/HVEM interactions for promoting anticancer T-cell responses. *PLoS One*, *12*(6), e0179201. https://doi.org/10.1371/journal.pone.0179201.

Stecher, C., Battin, C., Leitner, J., Zettl, M., Grabmeier-Pfistershammer, K., Holler, C., et al. (2017). PD-1 blockade promotes emerging checkpoint inhibitors in enhancing T cell responses to allogeneic dendritic cells. *Frontiers in Immunology*, *8*, 572. https://doi.org/10.3389/fimmu.2017.00572.

Steinberg, M. W., Huang, Y., Wang-Zhu, Y., Ware, C. F., Cheroutre, H., & Kronenberg, M. (2013). BTLA interaction with HVEM expressed on CD8(+) T cells promotes survival and memory generation in response to a bacterial infection. *PLoS One*, *8*(10), e77992. https://doi.org/10.1371/journal.pone.0077992.

Swerdlow, S. H., & International Agency for Research on Cancer & World Health Organization (2008). *WHO classification of tumours of haematopoietic and lymphoid tissues*. International Agency for Research on Cancer.

Tang, J., Pearce, L., O'Donnell-Tormey, J., & Hubbard-Lucey, V. M. (2018). Trends in the global immuno-oncology landscape. *Nature Reviews. Drug Discovery*, *17*, 783–784. https://doi.org/10.1038/nrd.2018.202.

Tang, H., Wang, Y., Chlewicki, L. K., Zhang, Y., Guo, J., Liang, W., et al. (2016). Facilitating T cell infiltration in tumor microenvironment overcomes resistance to PD-L1 blockade. *Cancer Cell*, *29*(3), 285–296. https://doi.org/10.1016/j.ccell.2016.02.004.

Tang, H., Zhu, M., Qiao, J., & Fu, Y. X. (2017). Lymphotoxin signalling in tertiary lymphoid structures and immunotherapy. *Cellular & Molecular Immunology*, *14*(10), 809–818. https://doi.org/10.1038/cmi.2017.13.

Taniguchi, K., & Karin, M. (2018). NF-kappaB, inflammation, immunity and cancer: Coming of age. *Nature Reviews. Immunology*, *18*(5), 309–324. https://doi.org/10.1038/nri.2017.142.

Tao, R., Wang, L., Murphy, K. M., Fraser, C. C., & Hancock, W. W. (2008). Regulatory T cell expression of herpesvirus entry mediator suppresses the function of B and T lymphocyte attenuator-positive effector T cells. *Journal of Immunology*, *180*(10), 6649–6655. 180/10/6649 [pii].

te Raa, G. D., Pascutti, M. F., Garcia-Vallejo, J. J., Reinen, E., Remmerswaal, E. B., ten Berge, I. J., et al. (2014). CMV-specific CD8 + T-cell function is not impaired in chronic lymphocytic leukemia. *Blood*, *123*(5), 717–724. https://doi.org/10.1182/blood-2013-08-518183.

Ten Hacken, E., Guieze, R., & Wu, C. J. (2017). SnapShot: Chronic lymphocytic leukemia. *Cancer Cell*, *32*(5). https://doi.org/10.1016/j.ccell.2017.10.015. 716-716.e17.

Thommen, D. S., Schreiner, J., Muller, P., Herzig, P., Roller, A., Belousov, A., et al. (2015). Progression of lung cancer is associated with increased dysfunction of T cells defined by coexpression of multiple inhibitory receptors. *Cancer Immunology Research*, *3*(12), 1344–1355. https://doi.org/10.1158/2326-6066.CIR-15-0097.

Thorsson, V., Gibbs, D. L., Brown, S. D., Wolf, D., Bortone, D. S., Ou Yang, T. H., et al. (2018). The immune landscape of cancer. *Immunity*, *48*(4). https://doi.org/10.1016/j.immuni.2018.03.023. 812–830.e14.

Tsang, J. Y. S., Chan, K. W., Ni, Y. B., Hlaing, T., Hu, J., Chan, S. K., et al. (2017). Expression and clinical significance of herpes virus entry mediator (HVEM) in breast Cancer. *Annals of Surgical Oncology*, *24*(13), 4042–4050. https://doi.org/10.1245/s10434-017-5924-1.

Tu, T. C., Brown, N. K., Kim, T. J., Wroblewska, J., Yang, X., Guo, X., et al. (2015). CD160 is essential for NK-mediated IFN-gamma production. *The Journal of Experimental Medicine*, *212*(3), 415–429. https://doi.org/10.1084/jem.20131601.

Vallabhapurapu, S., Matsuzawa, A., Zhang, W., Tseng, P. H., Keats, J. J., Wang, H., et al. (2008). Nonredundant and complementary functions of TRAF2 and TRAF3 in a ubiquitination cascade that activates NIK-dependent alternative NF-kappaB signaling. *Nature Immunology, 9*(12), 1364–1370. https://doi.org/10.1038/ni.1678.

Wang, B., Jie, Z., Joo, D., Ordureau, A., Liu, P., Gan, W., et al. (2017). TRAF2 and OTUD7B govern a ubiquitin-dependent switch that regulates mTORC2 signalling. *Nature, 545*(7654), 365–369. https://doi.org/10.1038/nature22344.

Wang, F. H., Wang, Y., Chen, Z. D., Chen, J. H., Qin, F. Z., Jiang, W. Q., et al. (2016). A phase IIa study of rhLTalpha-Da in combination with cisplatin and fluorouracil for patients with metastatic esophageal squamous cell carcinoma or gastric adenocarcinoma. *Medical Oncology, 33*(11), 125. https://doi.org/10.1007/s12032-016-0846-5.

Ward-Kavanagh, L. K., Lin, W. W., Sedy, J. R., & Ware, C. F. (2016). The TNF receptor superfamily in co-stimulating and co-inhibitory responses. *Immunity, 44*(5), 1005–1019. https://doi.org/10.1016/j.immuni.2016.04.019.

Ware, C. F. (2005). Network communications: Lymphotoxins, LIGHT, and TNF. *Annual Review of Immunology, 23*, 787–819.

Ware, C. F., & Sedy, J. R. (2011). TNF superfamily networks: Bidirectional and interference pathways of the herpesvirus entry mediator (TNFSF14). *Current Opinion in Immunology, 23*(5), 627–631. https://doi.org/10.1016/j.coi.2011.08.008.

Welsh, K., Milutinovic, S., Ardecky, R. J., Gonzalez-Lopez, M., Ganji, S. R., Teriete, P., et al. (2016). Characterization of potent SMAC mimetics that sensitize cancer cells to TNF family-induced apoptosis. *PLoS One, 11*(9), e0161952. https://doi.org/10.1371/journal.pone.0161952.

Xu, Z., Shi, R., Zhang, R., Zhang, D., & Wang, L. (2013). Association between tumor necrosis factor beta 252 a/G polymorphism and risk of gastric cancer: A meta-analysis. *Tumour Biology, 34*(6), 4001–4005. https://doi.org/10.1007/s13277-013-0989-3.

Yan, L., Da Silva, D. M., Verma, B., Gray, A., Brand, H. E., Skeate, J. G., et al. (2015). Forced LIGHT expression in prostate tumors overcomes Treg mediated immunosuppression and synergizes with a prostate tumor therapeutic vaccine by recruiting effector T lymphocytes. *Prostate, 75*(3), 280–291. https://doi.org/10.1002/pros.22914.

Yang, L., Feng, R., Liu, G., Liao, M., Zhang, L., & Wang, W. (2013). TNF-beta +252 A > G polymorphism and susceptibility to cancer. *Journal of Cancer Research and Clinical Oncology, 139*(5), 765–772. https://doi.org/10.1007/s00432-013-1384-6.

Ye, H., Park, Y. C., Kreishman, M., Kieff, E., & Wu, H. (1999). The structural basis for the recognition of diverse receptor sequences by TRAF2. *Molecular Cell, 4*(3), 321–330.

Yu, X., Huang, Y., Li, C., Yang, H., Lu, C., & Duan, S. (2014). Positive association between lymphotoxin-alpha variation rs909253 and cancer risk: A meta-analysis based on 36 case-control studies. *Tumour Biology, 35*(3), 1973–1983. https://doi.org/10.1007/s13277-013-1263-4.

Yu, P., Lee, Y., Liu, W., Chin, R. K., Wang, J., Wang, Y., et al. (2004). Priming of naive T cells inside tumors leads to eradication of established tumors. *Nature Immunology, 5*(2), 141–149.

Zelle-Rieser, C., Thangavadivel, S., Biedermann, R., Brunner, A., Stoitzner, P., Willenbacher, E., et al. (2016). T cells in multiple myeloma display features of exhaustion and senescence at the tumor site. *Journal of Hematology & Oncology, 9*(1), 116. https://doi.org/10.1186/s13045-016-0345-3.

Zhang, J., Baran, J., Cros, A., Guberman, J. M., Haider, S., Hsu, J., et al. (2011). International cancer genome consortium data portal—A one-stop shop for cancer genomics data. *Database: The Journal of Biological Databases and Curation, 2011*, bar026. https://doi.org/10.1093/database/bar026.

Zhang, B., Calado, D. P., Wang, Z., Frohler, S., Kochert, K., Qian, Y., et al. (2015). An oncogenic role for alternative NF-kappaB signaling in DLBCL revealed upon deregulated BCL6 expression. *Cell Reports*, *11*(5), 715–726. https://doi.org/10.1016/j.celrep.2015.03.059.

Zhang, Z. H., Chen, S., Chen, S., Liu, Y., & Qu, J. H. (2015). High levels of CD160 expression up-regulated counts of chronic lymphocytic leukemia cells and were associated with other clinical parameters in Chinese patients with chronic lymphocytic leukemia. *Leukemia & Lymphoma*, *56*(2), 529–532. https://doi.org/10.3109/10428194.2014.926345.

Zhang, J., Dominguez-Sola, D., Hussein, S., Lee, J. E., Holmes, A. B., Bansal, M., et al. (2015). Disruption of KMT2D perturbs germinal center B cell development and promotes lymphomagenesis. *Nature Medicine*, *21*(10), 1190–1198. https://doi.org/10.1038/nm.3940.

Zhang, T., Ye, L., Han, L., He, Q., & Zhu, J. (2016). Knockdown of HVEM, a lymphocyte regulator gene, in ovarian cancer cells increases sensitivity to activated T cells. *Oncology Research*, *24*(3), 189–196. https://doi.org/10.3727/096504016X14641336229602.

Zhao, Q., Huang, Z. L., He, M., Gao, Z., & Kuang, D. M. (2016). BTLA identifies dysfunctional PD-1-expressing CD4(+) T cells in human hepatocellular carcinoma. *Oncoimmunology*, *5*(12), e1254855. https://doi.org/10.1080/2162402X.2016.1254855.

Zhu, Y., Yao, S., Augustine, M. M., Xu, H., Wang, J., Sun, J., et al. (2016). Neuron-specific SALM5 limits inflammation in the CNS via its interaction with HVEM. *Science Advances*, *2*(4), e1500637. https://doi.org/10.1126/sciadv.1500637.

CHAPTER SIX

Pharmacology of ME-344, a novel cytotoxic isoflavone

Leilei Zhang[a,†], Jie Zhang[a,†], Zhiwei Ye[a], Danyelle M. Townsend[b], Kenneth D. Tew[a,*]

[a]Department of Cell and Molecular Pharmacology and Experimental Therapeutics, Medical University of South Carolina, Charleston, SC, United States
[b]Drug Discovery and Biomedical Sciences, Medical University of South Carolina, Charleston, SC, United States
*Corresponding author: e-mail address: tewk@musc.edu

Contents

1. Introduction — 188
2. Mitochondria as drug target organelles in cancer — 190
3. Cell components targeted by ME-344 — 191
   3.1 Mitochondrial enzyme complexes — 192
   3.2 mTOR inhibition through AMPK activation — 195
   3.3 ERK activation through increased ROS — 196
   3.4 Tubulin — 197
4. Preclinical investigational new drug (IND) enabling studies for ME-344 — 198
5. Clinical studies — 200
   5.1 Phase I: ME-344 as a single agent — 200
   5.2 Phase Ib: ME-344 in combination with topotecan — 202
6. Conclusions and future perspectives — 203
Acknowledgments — 203
Conflict of interest — 204
References — 204

Abstract

Isoflavones isolated from members of the *Fabaceae* (primarily *Leguminosae*) family have been characterized for their phytoestrogenic properties, but certain derivatives have also shown potential as possible cancer therapeutic agents. ME-344, related to phenoxodiol (Fig. 1), is a second generation isoflavone with a recent history of both preclinical and early clinical testing. The drug has unusual cytotoxicity profiles, where cancer cell lines can be categorized as either intrinsically sensitive or resistant to the

---

[†] Contributed equally.

drug. Evolving studies show that the cytotoxic properties of the drug are enacted through targeting mitochondrial bioenergetics. While the drug has undergone early Phase I/II trials in solid tumors with confined dose limiting effects and some evidence of disease response, there is a continuing need to define specific cellular targets that determine sensitivity, with the long-term goal of applying such information to individualized therapy. This review article details some of the existing and ongoing studies that are assisting in the continued drug development processes that may lead to new drug application (NDA) status.

1. Introduction

Among many adaptive changes, altered responses to caspase-dependent programmed cell death have been identified as a contributory cause of resistance to anticancer drugs (Igney & Krammer, 2002). Various isoflavones have been recognized in both cancer prevention (Sarkar & Li, 2003) and as inducers of cell death via caspase-independent pathways in drug resistant cancer cell lines. As such, there has been an emphasis on identifying those structures that may possess cytotoxic properties that can translate into a beneficial therapeutic index in cancer patients (Mor, Fu, & Alvero, 2006). Most isoflavones have been purified as natural plant compounds produced by members of the *Leguminous* family and can act as phytoestrogens with structures that bear resemblance to 17-β-estradiol (Yu et al., 2000). They share the same biosynthetic pathways used for flavonoid biosynthesis (Barnes, 2010), which produce active isoflavones such as daidzein and genistein. Glycoside conjugated forms of isoflavones frequently embody highly polar, water-soluble properties. Aglycone forms can be produced by intestinal bacteria, and these can have better absorption and biological activities than glycosidic forms (Vitale, Piazza, Melilli, Drago, & Salomone, 2013). Soybeans are the most significant dietary sources of isoflavones, with food processing conditions impacting their chemical structures. Genistein and daidzein have been extensively studied in clinical chemoprevention trials, particularly breast and prostate. Other therapeutic activities have been identified in cardiovascular disease, diabetes and post-menopausal disorders (Prasad, Phromnoi, Yadav, Chaturvedi, & Aggarwal, 2010; Setchell, 1998). A number of chemical activities have been ascribed to isoflavones, but in general, common to their mechanisms are effects upon redox homeostasis. These can then impact downstream events including protein tyrosine kinase (PTK) inhibition and deactivation of oncogenic proteins by catalysis of tyrosine phosphorylation. Inhibition of PTKs suppresses downstream oncogenic signaling pathways. Genistein and daidzein can also downregulate the anti-apoptotic protein B-cell lymphoma 2 (Bcl-2), upregulate the pro-apoptotic protein Bcl-2 associated X protein (Bax) and induce cell cycle G2/M arrest (Piontek, Hengels, Porschen, & Strohmeyer, 1993). In addition, genistein inhibits DNA topoisomerase activity and acts as a gene regulator (Kurzer & Xu, 1997).

NV-128 is a synthetic derivative of natural isoflavone compounds and has been investigated for its pre-clinical anti-cancer activities (Alvero et al., 2009, 2011). This agent is an analogue of phenoxodiol, a derivative of genistein. In contrast to phenoxodiol, NV-128 has been shown to induce caspase-independent cell death. Extrapolating from these compounds, a second generation isoflavone, ME-344, structurally similar to NV-128, but with greater anticancer potency in resistant cancer cells and stem cells has been developed. ME-344 has been evaluated in clinical situations (Bendell et al., 2015; Diamond et al., 2017), but there are numerous ongoing efforts to define how ME-344 mediates its cytotoxic properties. In this review, we provide an update of the pre-clinical and clinical data for ME-344 and provide perspectives on how this agent may move through development toward eventual potential FDA approval (Fig. 1).

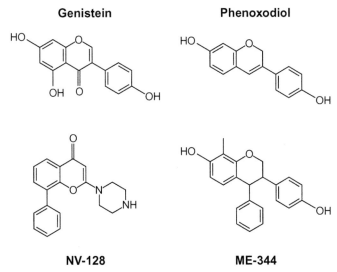

Fig. 1 Structures of common isoflavone molecules.

2. Mitochondria as drug target organelles in cancer

Mitochondria, the double-membrane-bound organelles in most eukaryotic cells, participate in bioenergetic metabolism producing ATP through electron transport chain (ETC) and oxidative phosphorylation (OXPHOS). Their relevance to ME-344 pharmacology resides in shared properties of overlapping ROS production, Ca^{2+} homeostasis and apoptosis initiation (Cheng & Ristow, 2013). Under physiological conditions, mitochondria are essential for cellular homeostasis, but during stress response, death pathways may be induced. Mitochondrial dysfunction has been implicated in many human diseases, including cancer (Fulda, Galluzzi, & Kroemer, 2010) and they have been targeted by different approaches in therapy, including: (i) Induction of mitochondrial permeability transition (MPT) leading to destruction of mitochondrial integrity. For example, Bcl-2 family members are critical regulators of MPT. Bcl-2 and Bcl-x_L proteins protect and stabilize mitochondrial membranes and prevent pro-apoptotic protein release into the cytosol. Conversely, Bax and Bcl-2 antagonist/killer (Bak) can create channels on outer mitochondrial membranes, inducing MPT causing cell death (Shamas-Din, Kale, Leber, & Andrews, 2013). (ii) Multiple copies of the mitochondrial genome (mtDNA), some of which code for OXPHOS complexes, can cause instability and can be targets. Mutations or aberrant copy numbers within the mtDNA create possible targets that may determine drug sensitivities (Singh et al., 1999). Either increased or decreased mtDNA copy number can occur in various solid tumors. In general, mtDNA loss in mammalian cells can be a precursor to cell death and the antibiotic ciprofloxacin can cause a disruption of mitochondrial function by cleaving mtDNA (Lawrence, Claire, Weissig, & Rowe, 1996). (iii) Enzymes in the OXPHOS pathway are also potential targets. OXPHOS is localized within inner mitochondrial membranes, composed of five mitochondrial enzyme complexes (complexes I–V) (Signes & Fernandez-Vizarra, 2018). The ETC is built with membrane channel pumps, which include complexes I–IV, involved in promoting redox reactions and pumping H^+ ions into mitochondrial intermembrane spaces. NADH and FADH from the Krebs Cycle act as electron carriers and fuel of the ETC. H^+ ions flow into the inter-membrane space of mitochondria to be channeled by electron acceptors and O_2, localized in complex IV, to produce water. These stepwise transfers lead to the serial pumping of protons into the mitochondrial inter-membrane space.

These protons are used for ATP production by ATP synthase (complex V). Mitochondrial OXPHOS is important in macromolecular biosynthesis, bioenergetics, but particularly in redox signaling and homeostasis (Chandel, 2015). By attenuating mitochondrial functions, deficiencies in OXPHOS have been implicated in drug resistance in cancer. Importantly in terms of cancer therapy, cancer cells, because of their high proliferative indexes, require and produce high levels of ATP and possess more active OXPHOS. Moreover, cancer stem cells also exhibit high levels of mitochondrial OXPHOS and generate more mitochondrial ROS (Viale et al., 2014). Thus, endogenously high levels of mitochondrial OXPHOS are seen in a variety of tumors (Roesch et al., 2013), creating a scenario where mitochondrial OXPHOS components may be considered as conceivable targets for anti-cancer drug development. Mitochondrial respiratory chain dysfunction, particularly involvement of complex I, is a contributory component in the pathogenesis of a variety of complex diseases including psychiatric disorders, neurodegenerative or cardiovascular diseases and cancer (Archer, 2013; Bansal & Kuhad, 2016; Bergman & Ben-Shachar, 2016; Kato, 2017; Looney & Childs, 1934). Mitochondrial complexes I and III have also been considered to be the two major enzyme complexes to generate ROS. Therefore, inhibition of mitochondrial complexes provides another perspective in treating human diseases. It is within this setting that ME-344 has gained credence as a cytotoxic anticancer drug.

3. Cell components targeted by ME-344

The biological effects of ME-344 have been reported to be quite broad, perhaps reflecting promiscuity of drug targeting. For example, ME-344 can activate caspase-dependent and independent cell death pathways, with implications to a number of plausible mechanisms of action. An analysis of the present literature allows us to deduce that these can include: inhibition of the respiratory complexes, i.e., complex I–V (Alvero et al., 2011; Lim, Carey, & McKenzie, 2015; Manevich, Reyes, Britten, Townsend, & Tew, 2016; Navarro et al., 2016) and translocation of HO-1 from rough endoplasmic reticulum to mitochondria (Manevich et al., 2016), accompanied by the downstream signaling events, e.g., (a) reduction in ATP production, activation of AMPK leading to mTOR-mediated autophagy, (b) increase in mitochondrial ROS production, activation of ERK leading to Bax-mediated loss of mitochondrial membrane potential (MMP, $\Delta\Psi_m$), endonuclease G (EndoG) translocation into nucleus which cleaves DNA, resulting in chromatin condensation. Possible inhibition of tubulin polymerization (Jeyaraju et al., 2016) has also been shown; however, it is not uncommon for estrogen-like molecules to have such effects (Speicher, Barone, & Tew, 1992; Speicher et al., 1994). Fig. 2 summarizes such information and the following sections provide further explanatory details.

3.1 Mitochondrial enzyme complexes

Complex I (NADH-coenzyme Q oxidoreductase) is the first enzyme complex of the ETC, reducing ubiquinone to ubiquinol by using the electrons donated from the oxidation of NADH from the Krebs cycle (Lenaz, Fato, Baracca, & Genova, 2004). It is also the largest multi-subunit enzyme of the ETC to generate an electrochemical proton gradient by pumping hydrogen ions across the inner mitochondrial membrane, driving ATP production and

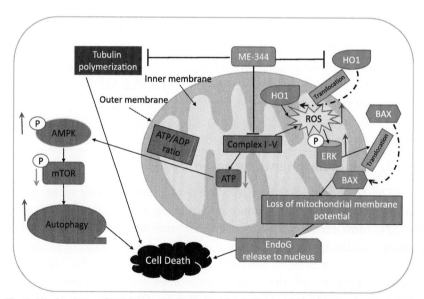

Fig. 2 Mechanisms of ME-344 action include (1) inhibition of the respiratory complexes, such as complex I–V; (2) inhibition of HO-1 and induction of its translocation to mitochondria, accompanied by downstream signaling events exemplified by, (a) reduction in ATP production, activation of AMPK leading to mTOR mediated autophagy, (b) increase in mitochondrial ROS production, activation of ERK leading to Bax-mediated loss of mitochondrial membrane potential, Endo G nuclease translocation into nucleus which cleaves DNA, resulting in chromatin condensation. Possible inhibition of tubulin polymerization has also been shown.

consequently the efficiency of the OXPHOS process (Holper, Ben-Shachar, & Mann, 2018; Yano, 2002). The impact of ME-344 on mitochondria was investigated by Lim et al., who showed that ME-344 suppresses mitochondrial respiration and complex I activity (and to some extent complex III). These effects were associated with a reduction in mitochondrial oxygen consumption and maximum ADP stimulated respiration rates. This inhibition has been compared to some of the characteristics of rotenone (Lim et al., 2015), a known complex I inhibitor, interfering with the ETC in mitochondria by preventing electron transfer from iron-sulfur centers in complex I to ubiquinone, subsequently inhibiting energy production and creating ROS by-products.

ME-344 also induced the dissipation of $\Delta\Psi_m$, comparable in extent to FCCP, a mitochondrial uncoupling agent that can depolarize mitochondrial membranes. This outcome was found in ovarian cancer stem cells treated with a related isoflavone, NV-128 (Fig. 1; Alvero et al., 2011; Lim et al., 2015). A normal $\Delta\Psi_m$ maintains steady-state levels of OXPHOS complexes, but in a variety of cancer cell lines and in ovarian cancer stem cells, ME-344 treatment was found to degrade the complex I subunit NDUFA9 and the complex IV subunits, COXI and IV. While such results support the principle that ME-344 interferes with mitochondrial respiration and dissipates $\Delta\Psi_m$, specific drug targets were not defined (Alvero et al., 2011; Lim et al., 2015).

In a cytotoxicity screen of 240 human tumor cell lines, 20 were shown to be intrinsically resistant to ME-344, while the remainder were sensitive. For our own studies with ME-344, we compared representative examples of both sensitive (SHP-77 and H460) and resistant (H596 and SW900) lung cancer cell lines, with the goal of identifying cellular traits that might be determinants of the mechanism of action of ME-344 (Manevich et al., 2016). Needless to say, such information should have relevance to how the drug is clinically applied. Consistent with other results which demonstrated ME-344 interference with mitochondrial energy production, we found that ME-344 caused substantial reduction in cellular *oxygen consumption rates (OCR)*, rapid hyperpolarization (instead of depolarization) in inner mitochondrial membranes, and oxidation of the protein thiols, with the effect more pronounced in sensitive lung cancer cell lines. In addition, ME-344 dose-dependently reduced ATP levels, ATP/ADP ratios, and increased the generation of H_2O_2 and $\cdot OH$ (not $O_2^{\cdot-}$) only in sensitive lung cancer cell lines, not in resistant lung cancer cell lines or in normal lung fibroblasts (MRC-5), implying a certain degree of specificity of ME-344 toward the sensitive cell lines. Furthermore, ME-344 induced a rapid increase and subsequent slow decrease of NADPH to below normal resting levels in sensitive, but not in resistant cancer or normal cell lines. Taken together, these results implicated several proteins and pathways that may be associated with the drug's mechanism of action. (1) ME-344 may target the oligomycin-sensitive F_0 component (proton pump) of ATP synthase (complex V), as evidenced by its ability to inhibit OCR and ATP production, inducing inner mitochondrial membrane hyperpolarization, and minimizing the effects of subsequent oligomycin in sensitive lung cancer cells. (2) ME-344 may have a direct impact on glycolysis. Most cancer cell lines depend on aerobic glycolysis for ATP synthesis and produce large amounts of lactate. As expected, initial extracellular acidification rate (ECAR) values of lung cancer cells were higher than those of normal cells. ME-344 has little effect on ECAR under glucose starvation. However, after addition of glucose, ME-344 significantly increased ECAR and glycolytic stress in sensitive lung cancer cells. (3) ME-344 influences heme oxygenase-1 (HO-1) expression, activity and subcellular localization. HO-1 is the rate-limiting enzyme involved in the oxidative degradation of free heme to produce biliverdin, carbon monoxide, and ferrous iron (Fe^{2+}) (Ryter, Alam, & Choi, 2006). HO-1 is generally regarded as cytoprotective; however, studies have shown that certain isoflavones can induce HO-1 expression leading either to protective events or to cause cell death (Park et al., 2011; Podkalicka, Mucha, Jozkowicz, Dulak, & Loboda, 2018), functional paradoxes perhaps linked with the mitochondrial translocation of HO-1 (Bansal, Biswas, & Avadhani, 2014). The cytotoxicity of ME-344 in sensitive lung cancer cells might be the consequence of increased ·OH production via the Fenton reaction, where the intermediate product of HO-1, Fe^{2+}, reacts with H_2O_2. ME-344 generated ROS imbues activities associated with a major redox transcription factor, nuclear factor, erythroid 2-like 2 (Nrf2), which regulates a series of antioxidant response genes (Tebay et al., 2015). In addition, the BTB and CNC Homology 1 (Bach1) protein can compete with Nrf2 for binding to antioxidant response elements (ARE), thereby suppressing some Nrf2 target genes, including heme oxygenase-1 (HO-1) (Reichard, Motz, & Puga, 2007; Shan, Lambrecht, Donohue, & Bonkovsky, 2006). Under basal conditions, Bach1 binds to ARE's that regulate HO-1 expression (Sun et al., 2002), but ROS can disrupt this and enable activation of HO-1 by Nrf2. These regulatory switches can influence the scope and extent of response to ME-344. (4) ME-344 may have an indirect (resulting from F_0-ATP synthase inhibition, inner mitochondrial membrane hyperpolarization and the compensatory proton fluxes) or direct impact on nicotinamide nucleotide transhydrogenase, an enzyme that uses energy from the mitochondrial proton gradient to inter-convert NADH to NADPH, as implied by the immediate effects of ME-344 on NADPH in sensitive lung cancer cells.

Each of these proteins or pathway targets is presently under continuing investigation, but our own ongoing studies (Zhang et al., 2019) have confirmed that ME-344 can bind to HO-1, inhibit its activity and initiate a robust Nrf2 signaling response to induce HO-1 which is accompanied by mitochondrial translocation of HO-1 in drug sensitive lung cancer cells. Similar effects were not seen in either drug resistant or normal lung fibroblasts cells, where basal levels of HO-1 were lower. This differential would be consistent with the reported ROS events and provide a rationale for an advantageous therapeutic index.

3.2 mTOR inhibition through AMPK activation

In a variety of sensitive cancer cell lines, ME-344 effects on mitochondrial complex subunits lead to significant loss of ATP, enhanced phosphorylation of AMPKα1, decreased phosphorylation of the mammalian target of rapamycin (mTOR; consisting of two distinct complexes, mTOR complex I (mTORC1) and mTOR complex II (mTORC2)) and decreased phosphorylation of p70 S6 kinase (Alvero et al., 2009, 2011; Bendell et al., 2015; Lim et al., 2015; Manevich et al., 2016). With low cellular ATP levels, AMP is more likely to activate AMPK by direct binding to its regulatory subunit (Oakhill et al., 2011; Xiao et al., 2011). One mechanism by which AMPK interacts with mTOR to regulate cell growth is by phosphorylating tumor suppressor tuberous sclerosis complex 2 (TSC2) (Mihaylova & Shaw, 2011). TSC1/2 gene mutation product TSC1/2 regulates cell growth negatively by acting upstream of mTOR and downstream of AMPK (Gwinn et al., 2008; Inoki et al., 2006). TSC2 accepts the signals and then regulates mTORC1 by interacting with the GTP-binding protein Ras homolog enriched in the brain (Rheb). GTPase-activating proteins (GAP) possessed by TSC2 increase the activity of Rheb to convert Rheb-GTP to Rheb-GDP which then inhibits the activation of mTORC1, with the end result that TSC2 suppresses the mTORC1 pathway (Gao & Pan, 2001; Kalender et al., 2010). ME-344 also decreases p-Akt, considered to have a role in induction of autophagy in cancer cells by interacting with mTOR. In addition, ME-344 increased the expression of mitochondrial beclin-1, which has been identified as a pro-autophagic molecule. Taken together, under energy starvation conditions, activation of AMPKα1 by ME-344 suppressed mTOR activity and since the mTOR pathway plays an integral role in regulating autophagy in eukaryotic cells (Yang, Liang, Gu, & Qin, 2005), increasing levels of LC-3 II caused by ME-344 treatment implied some cause:effect relationship with autophagy pathways in cancer cells (Alvero et al., 2009, 2011; Bendell et al., 2015). Nevertheless, these sorts of studies although informative did not identify direct drug binding targets.

3.3 ERK activation through increased ROS

Whenever drugs impact mitochondria and energy metabolism, there is a likelihood that cellular levels of ROS will be adversely impacted, resulting in excessive damage to a variety of biomolecules in the inner mitochondrial membrane and matrix (Adam-Vizi & Chinopoulos, 2006). ME-344 was shown to increase the levels of mitochondrial superoxide and cellular hydroxyl radicals and hydrogen peroxide (H_2O_2), but primarily (although not exclusively) in drug-sensitive cancer cells and cancer stem cells (Alvero et al., 2011; Bendell et al., 2015). ME-344 induced inhibition of OCR may also be responsible for oxidative stress generated by impairment of mitochondrial ETC. Characteristic of most cancer cells is aberrations and inefficiencies in OXPHOS and accompanying antioxidant systems and even under cellular homeostasis, cancer cells produce high levels of ROS and have lower levels of antioxidant enzymes (Mates & Sanchez-Jimenez, 1999). Since intermediate signaling events can be regulated by ROS, instead of inducing cell death, these molecules can alter tumor growth. Alvero et al. 2011 showed that ME-344, by inhibiting ETC, can increase mitochondrial ROS, leading to the activation of mitochondrial ERK1/2 in cancer stem cells. Consistently, pre-treatment with MnTBAP, a cell permeable superoxide dismutase mimetic, diminishes ME-344-induced activation of ERK1/2, indicating that ERK activation in cancer stem cells is linked with increasing mitochondrial ROS (Alvero et al., 2011; Bendell et al., 2015). Different subcellular localization of MEK/ERK signaling pathways can regulate either cell death or survival (Alavi, Hood, Frausto, Stupack, & Cheresh, 2003; Chu, Levinthal, Kulich, Chalovich, & DeFranco, 2004). For example, activation can cause ERK1/2 nuclear translocation and subsequent activation of transcription factors leading to cell survival (Seger & Krebs, 1995). In contrast, mitochondrial ERK can participate in cell death pathways by regulating various mitochondrial functions. In Parkinson's disease, ERK activity has been linked with oxidative injury through initiation of mitochondrial ROS (Kulich, Horbinski, Patel, & Chu, 2007). ME-344 can induce a pro-apoptotic effect in cancer stem cells through upregulation of the Bcl-2 family member, Bax. Pre-treatment with the ERK inhibitor U0126 can abrogate the effects of ME-344 on Bax mitochondrial translocation (Alvero et al., 2011, Bendell et al., 2015). Bcl-2 family members play a role in mitochondrial targeting of MEK/ERK signaling pathways, where for example, Bax, as a pro-apoptotic member, is expressed in the cytosol of healthy cells. Under stress conditions, Bcl-2 is phosphorylated by ERK signaling leading to Bax translocation from cytosol to mitochondria by integrating into the outer mitochondrial membrane via its carboxyl terminus (Mikhailov et al., 2001; Nechushtan, Smith, Hsu, & Youle, 1999). ME-344 induces MPT, caused when Bax forms channels in the outer mitochondrial membrane in conjunction with ME-344 induced dissipation of $\Delta\Psi_m$. ME-344 initiated beclin-1 mitochondrial translocation might also contribute to the MPT, since mitochondrial integrity is regulated to some degree by Bcl-2 family members. Under normal conditions, Bcl-2 and Bcl-x_L reside in mitochondria and stabilize membrane structure. ME-344 induced mitochondrial translocation of beclin-1 allows it to bind to Bcl-2 and Bcl-x_L through its BH domain and abrogates the inhibition of Bax by Bcl-2, leading to Bax induced MPT. Therefore, ME-344-induced ERK signaling activation can drive cellular loss of $\Delta\Psi_m$ via the interaction between Bax and mitochondrial components (Alvero et al., 2011, Bendell et al., 2015). ME-344 also induces EndoG nuclear translocation, resulting in chromatin condensation. ME-344 induced the loss of $\Delta\Psi_m$ and MPT caused release of some mitochondrial proteins, including the mitochondrial nucleases EndoG and AIF. Nuclear translocation of EndoG, not AIF, can contribute to the cleavage of DNA and chromatin condensation effects initiated by ME-344 (Alvero et al., 2009; Bendell et al., 2015).

3.4 Tubulin

ME-344 was shown to cause morphological changes in the plasma membranes of acute myeloid leukemia (AML) cells, effects that were considered to be similar to vinblastine, a characterized tubulin inhibitor (Jeyaraju et al., 2016). Indeed, ME-344 was shown to inhibit tubulin polymerization in AML cells by interacting at the colchicine binding site. Consequently, ME-344 caused specific G2/M cell cycle arrest by suppression of microtubule dynamics and demonstrated synergistic cell killing effects with vinblastine. Because estrogenic molecules are known to have antimicrotubule effects (Speicher et al., 1992, 1994), this result may reflect the phytoestrogenic properties (Leclercq & Jacquot, 2014) of the isoflavone backbone. Interestingly, combinations of such molecules have been previously tested in prostate cancer patients (Hudes et al., 1992), although the actual specificities of this tubulin binding and its importance, at low concentrations, to the in vivo or clinical effects remain to be established.

4. Preclinical investigational new drug (IND) enabling studies for ME-344

Characteristics of many tumors are high glycolytic rates, abnormal tumor angiogenesis and hypoxia. Some antiangiogenic agents can correct pathologic tumor angiogenesis and normalize tumor hypoxia (Jain, 2013; Kerbel, 2006). Antiangiogenics are widely used in the treatment of tumors such as kidney, liver, breast, ovarian, colorectal and lung. However, all patients receiving these biological agents eventually experience acquired resistance to the drugs and subsequent progressive disease, frequently as a consequence of upregulation of aerobic glycolysis mediated by mutations in MAPK and PI3K/AKT pathways (Mastri et al., 2016; Quintela-Fandino, 2016; Sennino & McDonald, 2012). Recent studies suggested that ME-344 could overcome acquired resistance to antiangiogenics such as multi-tyrosine-kinase inhibitor (TKIs) (Navarro et al., 2016). TKI's (nintedanib or dovitinib) caused 60% tumor growth inhibition (TGI), accompanied by a twofold decrease in glucose uptake and reduced glycolysis through decreased HIF1α and AKT signaling, as evidenced by diminished glycolysis metabolites and end products of the pentose-phosphate shunt. How tumors were able to continue growing under such nutritional stress conditions was the subject of this study by Quintela-Fandino's group (Navarro et al., 2016). This suggested that tumor metabolism can be reprogrammed in response to TKI or decreased glycolysis, with mitochondrial metabolism increasing under the control of PPAR-α, AMPK and PKA activation. Evidence of increased mitochondrial metabolism in TKI- versus vehicle-treated tumors included: (1) an ∼twofold increase in ketones and fatty acids, substrates for the tricarboxylic acid cycle, was detected in tumors treated with TKI. (2) A >2.5-fold upregulation of the ketone-body transporters MCT1 and 4 was detected in TKI-treated tumors, coupled with a >fourfold increase in the levels of ACAT1, the enzyme that allows re-use of ketone bodies as mitochondrial substrates. (3) Adipose deposits almost disappeared with TKI treatment, coupled with increased levels of the fatty acid transporters FABP3, CAV1, and SLC27A1/A2 and the lipases PNPLA2 and 3, and an almost fourfold upregulation of CPT1B, which shuttles the carnitine-bound species committed for mitochondrial degradation into the mitochondrial matrix. (4) Succinate dehydrogenase (SDH) or complex II activity increased from week 1 of TKI treatment and was maintained for >2 months. (5) OCR was ~twofold higher in tumors treated with TKI. These adaptive changes in energy metabolism were mechanistic adaptations that permitted resistance to chronic TKI treatment and indicated that pharmacologic modulation of mitochondrial respiration might not be efficient if used as monotherapy, but may enhance the effects of the TKIs. When one energy source (glycolysis) is pharmacologically restricted, tumors become vulnerable to inhibition of the other (mitochondrial metabolism). Indeed, while neither ME-344 nor phenformine was effective when administered as monotherapy, the combination demonstrated synergistic effects with TKI (nintedanib, dovitinib or regorafenib) through inhibition of both mitochondrial and glycolytic metabolisms. The therapeutic benefit of such a drug combination has been successfully extended to two lung cancer models: Pulm24 lung-cancer-patient-derived xenograft and spontaneous Lewis lung carcinoma cancer model. This approach certainly appears to hold future promise and partly as a consequence of the broad spectrum of cytotoxic activity of ME-344 in human tumor cell lines (including a number of drug resistant variants) led to its application in reversing TKI resistance in a spontaneous HER-2 negative (estrogen receptor positive, and progesterone receptor negative) breast cancer model (Navarro et al., 2016). After priming with TKI antiangiogenics, mitochondrial metabolism, as compensation for reduced glycolysis, becomes essential for tumor survival and can lead to acquired resistance against these drugs. For example, nintedanib, in combination with ME-344, demonstrated synergistic effects, increasing the TGI from 64% to 92%. Effective therapeutic synergy was also observed with ME-344 and regorafenib (another TKI), increasing the TGI to 88% (Navarro et al., 2016). On the basis of these findings, an early-stage clinical trial that combines antiangiogenic treatment (bevacizumab, 15 mg/kg at day 1) with weekly doses of ME-344 (10 mg/kg on days 8, 15 and 22) has been launched for patients with early HER-2 negative breast cancer. So far, the authors claim that interim data from 19 patients are encouraging and suggest that completion of enrollment of the clinical study of ME-344 in combination with bevacizumab is warranted (Miguel Quintela-Fandino et al., 2018).

AML accounts for 10–15% of all leukemias diagnosed in children and is the most common acute leukemia in adults. With AML, the incidence escalates with age and those >60 have poorer prognosis due to toxicity and low response to standard chemotherapies. AML is not only a highly aggressive disease, but also develops resistance to further chemotherapy (Estey & Dohner, 2006; Siegel, Naishadham, & Jemal, 2013). Studies have been carried out to investigate the cytotoxic effect of ME-344 on leukemia cell lines, primary AML patient samples and normal hematopoietic cells (Jeyaraju et al., 2016). ME-344 treatment resulted in inhibition of cell growth and viability in all the leukemia cell lines tested, e.g., OCI-AML2, TEX, HL60, K562, KG1a, U937 and NB4, with IC_{50} values in the range of 70–260nM. In addition, ME-344 was toxic to AML patient samples rather than normal hematopoietic samples. Furthermore, in leukemia xenograft models using OCI-AML2 or MDAY-D2, ME-344 reduced tumor growth by up to 95% of control without evidence of toxicity. Taken together, these results implied that ME-344 has anti-leukemia efficacy both in vitro and in vivo, which supports its clinical evaluation in patients with AML and other hematological malignancies.

5. Clinical studies

Results from Phase I and Ib trials have been published. ME-344 has been employed to treat patients with locally advanced or metastatic refractory solid tumors by single agent or combination therapies. The original rationale for such trials was the preclinical data showing that activation of caspase-independent cell death pathways in malignant cells may create a beneficial therapeutic index. Two clinical trials with ME-344 as a single agent or in combination with topotecan (Table 1) have been completed in patients with refractory solid tumors and were designed to show safety, adverse events, pharmacokinetics and disease response (Bendell et al., 2015; Diamond et al., 2017).

5.1 Phase I: ME-344 as a single agent

The first-in-human phase 1 study of ME-344 identified safety and adverse events (AEs) that might provide dose-limiting toxicities (DLTs) and maximum tolerated dose (MTD). Thirty patients with refractory solid tumors received intravenous ME-344 at six dose levels (1.25–20mg/kg) on days 1, 8, 15, and 22 of each 28-day treatment cycle until disease progression. DLTs developing at doses 15 and 20mg/kg included grade 3 neuropathy,

Table 1 Summary of clinical trials with ME-344.

Clinical trials	Patients/type of cancer, no. (%)	Median no. of prior therapies, no. (%)	Pharmacokinetics	Adverse events, no. (%)	Response, no. (%)
Phase I	30 patients: 5 (16.7) NSCLC; 5 (16.7) colorectal; 3 (10.0) endometrial; 2 (6.7) ovarian; 2 (6.7) squamous cell carcinoma of vagina; 2 (6.7) urothelial carcinoma; 11 (36.7) other[a]	3 (10.0) 0–1 prior therapies; 5 (16.7) 2 prior therapies; 8 (26.7) 3 prior therapies; 14 (46.7) 4 prior therapies	C_{max}/dose ($\mu g/mL/mg$): 1.99; $t_{1/2}$ (h): 6; AUC/dose ($h \times \mu g/mL/mg$): 0.008	6 (20.0) neuropathy; 6 (20.0) nausea; 6 (20.0) dizziness; 5 (16.7) fatigue; 4 (13.3) vomiting; 3 (10.0) diarrhea; 3 (10.0) asthenia	1 (3.3) partial response; 10 (33.3) stable disease; 10 (33.3) progressive disease; 9 (30.0) not evaluable
Phase Ib	46 patients: 28 (60.9) ovarian; 13 (28.3) SCLC; 5 (10.9) cervical	28 (60.9) 1–4 prior therapies in ovarian cancer; 13 (28.3) 1–3 prior therapies in SCLC; 5 (10.9) 1–4 prior therapies in cervical cancer	C_{max} (ng/mL): 20,880; C_{min} (ng/mL): 25.30; t_{max} (h): 0.500; AUC_{0-t} ($h \times ng/mL$): 21,830; AUC_{0-inf} ($h \times ng/mL$): 22,040; $AUC_{\%\,extrap}$ (%): 1.03; $t_{1/2}$ (h): 5.30; k_{el} (1/h): 0.14; $CL_{,ss}$ (L/h): 42.59; $V_{d,ss}$ (L): 104.65; $CL_{,ss}/kg$ (L/h): 0.50; $V_{d,ss}/kg$ (L): 1.20	30 (65.2) fatigue; 26 (56.5) neutropenia; 23 (50.0) thrombocytopenia; 22 (47.8) nausea; 21 (45.7) diarrhea; 19 (41.3) decreased appetite; 19 (41.3) hypertension; 18 (39.1) vomiting; 16 (34.8) anemia; 15 (32.6) constipation; 9 (19.6) weight decreased; 8 (17.4) arthralgia; 8 (17.4) back pain; 8 (17.4) dyspnea; 7 (15.2) abdominal pain; 7 (15.2) asthenia; 7 (15.2) cough	1 (2.4) partial response; 21 (51.2) stable disease; 9 (22.0) progressive disease; 10 (24.4) not evaluable

[a] Other includes bladder, breast, carcinoid of the ileum, cervical, cervical leiomyosarcoma, small cell lung, melanoma, pancreatic, peritoneal, sarcoma, and unknown primary (one patient each).
NSCLC, Non-small cell lung carcinoma; SCLC, Small cell lung carcinoma.

vomiting, nausea and hypertension, but each reverted to normal with discontinuation of treatment. Therefore, in subsequent studies 10 mg/kg was selected to be the MTD. Common treatment-related AEs included neuropathy (20.0%), nausea (20.0%), dizziness (20.0%), fatigue (16.7%), vomiting (13.3%), diarrhea (10.0%) and asthenia (10.0%) (Table 1). 6 of 30 patients discontinued treatments as a consequence of serious AEs (SAE). Pharmacokinetic parameters were assessed on day 1 and day 15. The mean terminal half-life of ME-344 was 6h; maximal plasma concentration C_{max}/dose was 1.99 µg/mL/mg; mean area under the concentration curve AUC/dose was 0.008 h × µg/mL/mg. There were no significant differences in volume of distribution at steady state (V_{dss}) and drug clearance (Cl). Efficacy was assessed in 30 patients. 21 of 30 received at least 3 cycles of treatment. One (3.3%) patient with small cell lung cancer (SCLC) experienced a partial response (PR) at the 11th cycle of treatment at a dose of 5 mg/kg. Laminography images showed decreased tumor bulk at week 52 compared to baseline. Within the cohort, 4 of 10 patients were reported to have stable disease (SD), including: urothelial carcinoma for 47 weeks; carcinoid of the ileum for 40 weeks; cervical leiomyosarcoma for 39 weeks; cervical cancer for 31 weeks. 10 patients (33.3%) developed progressive disease (PD) within the treatment duration. Overall, ME-344 was judged to have demonstrated promising efficacy as a monotherapy in this phase I study and provided a baseline for future combination studies.

5.2 Phase Ib: ME-344 in combination with topotecan

A subsequent phase Ib study was implemented to evaluate ME-344 in combination with topotecan in patients with SCLC, ovarian or cervical cancer. Topotecan is an established drug in the management of these diseases (Hirte, Kennedy, Elit, & Fung Kee Fung, 2015; Riemsma, Simons, Bashir, Gooch, & Kleijnen, 2010). 46 patients (32 of whom had been previously treated) received ME-344 intravenously at 10 mg/kg on days 1, 8, 15, and 22, with topotecan at 4 mg/m² on days 1, 8, and 15 of each 28-day treatment cycle. Discontinuation was enacted if disease progression occurred (Diamond et al., 2017; Safra et al., 2013; Sehouli et al., 2011). Common treatment-related AEs were similar to the other trial, including fatigue (65.2%), neutropenia (56.5%), thrombocytopenia (50.0%), nausea (47.8%), diarrhea (45.7%), decreased appetite (41.3%) and hypertension (41.3%) (Table 1). Only two patients (4.3%) experienced neuropathy with no grade 3/4 AEs. 3 patients (6.5%) who developed ME-344-related AEs discontinued treatment. One patient (2.1%) experienced 1 ME-344-related grade 3 diarrhea; 7 patients (15.2%) experienced 6 topotecan-related SAEs, including grade 4 thrombocytopenia, grade 3 fatigue, grade 3 febrile neutropenia, grade 3 blood-stream infection and neutropenic sepsis. 5 patients (10.9%) died within 30 days of discontinuing treatment. Pharmacokinetic parameters were similar to the earlier trial and decreasing plasma levels of ME-344 suggested little drug accumulation. One patient (2.4%) with ovarian cancer obtained a PR and continued for 14 cycles. 21 patients (51.2%) experienced SD, translating into a clinical benefit of 53.7% and an overall response rate of 2.4%. These results were unspectacular, but did not preclude the possibility that different combinations should be explored (Diamond et al., 2017). As discussed in a previous section, the combination of ME-344 with antiangiogenic agents has become the next-line clinical approach.

6. Conclusions and future perspectives

The current literature contains significant preclinical data and more limited clinical results that support the continued development of ME-344 as a cytotoxic anticancer drug. Isoflavones have been the subject of numerous interrogations as cancer preventative agents. However, the phenoxodiol scaffold of ME-344 has cytotoxic properties that have been transmitted into the general pharmacological characteristics of ME-344. Screening of numerous cancer cell lines showed that cells were either intrinsically sensitive or resistant to the drug. Evidence so far is that energy metabolism and mitochondria constitute targets for the drug, although precisely which mitochondrial proteins are specifically inhibited remains under study. Delineation of such (a) target(s) will assist significantly in moving drug development forward, perhaps predicting which tumor characteristics may predetermine drug response. In the future, such information may be applied to a precision medicine or individualized therapy setting, where biomarkers of drug sensitivity may be predicted.

Acknowledgments

This work was supported by grants from the National Institutes of Health (5P20GM103542, COBRE in Oxidants, Redox Balance and Stress Signaling), support from the South Carolina Centers of Excellence program and was conducted in a facility constructed with the support from the National Institutes of Health, Grant Number C06 RR015455 from the Extramural Research Facilities Program of the National Center for Research Resources. Financial support from MEI Pharma is also acknowledged.

Conflict of interest

K.D.T. has been the recipient of research support from MEI Pharm

References

Adam-Vizi, V., & Chinopoulos, C. (2006). Bioenergetics and the formation of mitochondrial reactive oxygen species. *Trends in Pharmacological Sciences, 27*(12), 639–645.

Alavi, A., Hood, J. D., Frausto, R., Stupack, D. G., & Cheresh, D. A. (2003). Role of Raf in vascular protection from distinct apoptotic stimuli. *Science, 301*(5629), 94–96.

Alvero, A. B., Montagna, M. K., Chen, R., Kim, K. H., Kyungjin, K., Visintin, I., et al. (2009). NV-128, a novel isoflavone derivative, induces caspase-independent cell death through the Akt/mammalian target of rapamycin pathway. *Cancer, 115*(14), 3204–3216.

Alvero, A. B., Montagna, M. K., Holmberg, J. C., Craveiro, V., Brown, D., & Mor, G. (2011). Targeting the mitochondria activates two independent cell death pathways in ovarian cancer stem cells. *Molecular Cancer Therapeutics, 10*(8), 1385–1393.

Archer, S. L. (2013). Mitochondrial dynamics—Mitochondrial fission and fusion in human diseases. *The New England Journal of Medicine, 369*(23), 2236–2251.

Bansal, S., Biswas, G., & Avadhani, N. G. (2014). Mitochondria-targeted heme oxygenase-1 induces oxidative stress and mitochondrial dysfunction in macrophages, kidney fibroblasts and in chronic alcohol hepatotoxicity. *Redox Biology, 2*, 273–283.

Bansal, Y., & Kuhad, A. (2016). Mitochondrial dysfunction in depression. *Current Neuropharmacology, 14*(6), 610–618.

Barnes, S. (2010). The biochemistry, chemistry and physiology of the isoflavones in soybeans and their food products. *Lymphatic Research and Biology, 8*(1), 89–98.

Bendell, J. C., Patel, M. R., Infante, J. R., Kurkjian, C. D., Jones, S. F., Pant, S., et al. (2015). Phase 1, open-label, dose escalation, safety, and pharmacokinetics study of ME-344 as a single agent in patients with refractory solid tumors. *Cancer, 121*(7), 1056–1063.

Bergman, O., & Ben-Shachar, D. (2016). Mitochondrial oxidative phosphorylation system (OXPHOS) deficits in schizophrenia: Possible interactions with cellular processes. *Canadian Journal of Psychiatry, 61*(8), 457–469.

Chandel, N. S. (2015). Evolution of mitochondria as signaling organelles. *Cell Metabolism, 22*(2), 204–206.

Cheng, Z., & Ristow, M. (2013). Mitochondria and metabolic homeostasis. *Antioxidants & Redox Signaling, 19*(3), 240–242.

Chu, C. T., Levinthal, D. J., Kulich, S. M., Chalovich, E. M., & DeFranco, D. B. (2004). Oxidative neuronal injury. The dark side of ERK1/2. *European Journal of Biochemistry, 271*(11), 2060–2066.

Diamond, J. R., Goff, B., Forster, M. D., Bendell, J. C., Britten, C. D., Gordon, M. S., et al. (2017). Phase Ib study of the mitochondrial inhibitor ME-344 plus topotecan in patients with previously treated, locally advanced or metastatic small cell lung, ovarian and cervical cancers. *Investigational New Drugs, 35*(5), 627–633.

Estey, E., & Dohner, H. (2006). Acute myeloid leukaemia. *Lancet, 368*(9550), 1894–1907.

Fulda, S., Galluzzi, L., & Kroemer, G. (2010). Targeting mitochondria for cancer therapy. *Nature Reviews. Drug Discovery, 9*(6), 447–464.

Gao, X., & Pan, D. (2001). TSC1 and TSC2 tumor suppressors antagonize insulin signaling in cell growth. *Genes & Development, 15*(11), 1383–1392.

Gwinn, D. M., Shackelford, D. B., Egan, D. F., Mihaylova, M. M., Mery, A., Vasquez, D. S., et al. (2008). AMPK phosphorylation of raptor mediates a metabolic checkpoint. *Molecular Cell, 30*(2), 214–226.

Hirte, H., Kennedy, E. B., Elit, L., & Fung Kee Fung, M. (2015). Systemic therapy for recurrent, persistent, or metastatic cervical cancer: A clinical practice guideline. *Current Oncology, 22*(3), 211–219.

Holper, L., Ben-Shachar, D., & Mann, J. J. (2018). Multivariate meta-analyses of mitochondrial complex I and IV in major depressive disorder, bipolar disorder, schizophrenia, Alzheimer disease, and Parkinson disease. *Neuropsychopharmacology*, 1–13.

Hudes, G. R., Greenberg, R., Krigel, R. L., Fox, S., Scher, R., Litwin, S., et al. (1992). Phase II study of estramustine and vinblastine, two microtubule inhibitors, in hormone-refractory prostate cancer. *Journal of Clinical Oncology, 10*(11), 1754–1761.

Igney, F. H., & Krammer, P. H. (2002). Death and anti-death: Tumour resistance to apoptosis. *Nature Reviews. Cancer, 2*(4), 277–288.

Inoki, K., Ouyang, H., Zhu, T., Lindvall, C., Wang, Y., Zhang, X., et al. (2006). TSC2 integrates Wnt and energy signals via a coordinated phosphorylation by AMPK and GSK3 to regulate cell growth. *Cell, 126*(5), 955–968.

Jain, R. K. (2013). Normalizing tumor microenvironment to treat cancer: Bench to bedside to biomarkers. *Journal of Clinical Oncology, 31*(17), 2205–2218.

Jeyaraju, D. V., Hurren, R., Wang, X., MacLean, N., Gronda, M., Shamas-Din, A., et al. (2016). A novel isoflavone, ME-344, targets the cytoskeleton in acute myeloid leukemia. *Oncotarget, 7*(31), 49777–49785.

Kalender, A., Selvaraj, A., Kim, S. Y., Gulati, P., Brule, S., Viollet, B., et al. (2010). Metformin, independent of AMPK, inhibits mTORC1 in a rag GTPase-dependent manner. *Cell Metabolism, 11*(5), 390–401.

Kato, T. (2017). Neurobiological basis of bipolar disorder: Mitochondrial dysfunction hypothesis and beyond. *Schizophrenia Research, 187*, 62–66.

Kerbel, R. S. (2006). Antiangiogenic therapy: A universal chemosensitization strategy for cancer? *Science, 312*(5777), 1171–1175.

Kulich, S. M., Horbinski, C., Patel, M., & Chu, C. T. (2007). 6-Hydroxydopamine induces mitochondrial ERK activation. *Free Radical Biology & Medicine, 43*(3), 372–383.

Kurzer, M. S., & Xu, X. (1997). Dietary phytoestrogens. *Annual Review of Nutrition, 17*, 353–381.

Lawrence, J. W., Claire, D. C., Weissig, V., & Rowe, T. C. (1996). Delayed cytotoxicity and cleavage of mitochondrial DNA in ciprofloxacin-treated mammalian cells. *Molecular Pharmacology, 50*(5), 1178–1188.

Leclercq, G., & Jacquot, Y. (2014). Interactions of isoflavones and other plant derived estrogens with estrogen receptors for prevention and treatment of breast cancer-considerations concerning related efficacy and safety. *The Journal of Steroid Biochemistry and Molecular Biology, 139*, 237–244.

Lenaz, G., Fato, R., Baracca, A., & Genova, M. L. (2004). Mitochondrial quinone reductases: Complex I. *Methods in Enzymology, 382*, 3–20.

Lim, S. C., Carey, K. T., & McKenzie, M. (2015). Anti-cancer analogues ME-143 and ME-344 exert toxicity by directly inhibiting mitochondrial NADH: Ubiquinone oxidoreductase (complex I). *American Journal of Cancer Research, 5*(2), 689–701.

Looney, J. M., & Childs, H. M. (1934). The lactic acid and glutathione content of the blood of schizophrenic patients. *The Journal of Clinical Investigation, 13*(6), 963–968.

Manevich, Y., Reyes, L., Britten, C. D., Townsend, D. M., & Tew, K. D. (2016). Redox signaling and bioenergetics influence lung cancer cell line sensitivity to the isoflavone ME-344. *The Journal of Pharmacology and Experimental Therapeutics, 358*(2), 199–208.

Mastri, M., Rosario, S., Tracz, A., Frink, R. E., Brekken, R. A., & Ebos, J. M. (2016). The challenges of modeling drug resistance to antiangiogenic therapy. *Current Drug Targets, 17*(15), 1747–1754.

Mates, J. M., & Sanchez-Jimenez, F. (1999). Antioxidant enzymes and their implications in pathophysiologic processes. *Frontiers in Bioscience, 4*, D339–D345.

Miguel Quintela-Fandino, J. V. A., Salgado, A. C., Mouron, S. A., Guerra, J. A., Cortes, M. G., Morente, M., et al. (2018). Abrogation of resistance against bevacizumab (Bev) by mitochondrial inhibition: A phase 0 randomized trial of Bev plus ME344 or placebo in early HER2-negative breast cancer (HERNEBC). *Journal of Clinical Oncology, 36*, 2552. abstract.

Mihaylova, M. M., & Shaw, R. J. (2011). The AMPK signalling pathway coordinates cell growth, autophagy and metabolism. *Nature Cell Biology, 13*(9), 1016–1023.

Mikhailov, V., Mikhailova, M., Pulkrabek, D. J., Dong, Z., Venkatachalam, M. A., & Saikumar, P. (2001). Bcl-2 prevents Bax oligomerization in the mitochondrial outer membrane. *The Journal of Biological Chemistry, 276*(21), 18361–18374.

Mor, G., Fu, H. H., & Alvero, A. B. (2006). Phenoxodiol, a novel approach for the treatment of ovarian cancer. *Current Opinion in Investigational Drugs, 7*(6), 542–548.

Navarro, P., Bueno, M. J., Zagorac, I., Mondejar, T., Sanchez, J., Mouron, S., et al. (2016). Targeting tumor mitochondrial metabolism overcomes resistance to antiangiogenics. *Cell Reports, 15*(12), 2705–2718.

Nechushtan, A., Smith, C. L., Hsu, Y. T., & Youle, R. J. (1999). Conformation of the Bax C-terminus regulates subcellular location and cell death. *The EMBO Journal, 18*(9), 2330–2341.

Oakhill, J. S., Steel, R., Chen, Z. P., Scott, J. W., Ling, N., Tam, S., et al. (2011). AMPK is a direct adenylate charge-regulated protein kinase. *Science, 332*(6036), 1433–1435.

Park, J. S., Jung, J. S., Jeong, Y. H., Hyun, J. W., Le, T. K., Kim, D. H., et al. (2011). Antioxidant mechanism of isoflavone metabolites in hydrogen peroxide-stimulated rat primary astrocytes: Critical role of hemeoxygenase-1 and NQO1 expression. *Journal of Neurochemistry, 119*(5), 909–919.

Piontek, M., Hengels, K. J., Porschen, R., & Strohmeyer, G. (1993). Antiproliferative effect of tyrosine kinase inhibitors in epidermal growth factor-stimulated growth of human gastric cancer cells. *Anticancer Research, 13*(6A), 2119–2123.

Podkalicka, P., Mucha, O., Jozkowicz, A., Dulak, J., & Loboda, A. (2018). Heme oxygenase inhibition in cancers: Possible tools and targets. *Contemporary Oncology (Poznań, Poland), 22*(1A), 23–32.

Prasad, S., Phromnoi, K., Yadav, V. R., Chaturvedi, M. M., & Aggarwal, B. B. (2010). Targeting inflammatory pathways by flavonoids for prevention and treatment of cancer. *Planta Medica, 76*(11), 1044–1063.

Quintela-Fandino, M. (2016). Normoxic or hypoxic adaptation in response to antiangiogenic therapy: Clinical implications. *Molecular & Cellular Oncology, 3*(5), e1217368.

Reichard, J. F., Motz, G. T., & Puga, A. (2007). Heme oxygenase-1 induction by NRF2 requires inactivation of the transcriptional repressor BACH1. *Nucleic Acids Research, 35*(21), 7074–7086.

Riemsma, R., Simons, J. P., Bashir, Z., Gooch, C. L., & Kleijnen, J. (2010). Systematic review of topotecan (hycamtin) in relapsed small cell lung cancer. *BMC Cancer, 10*, 436.

Roesch, A., Vultur, A., Bogeski, I., Wang, H., Zimmermann, K. M., Speicher, D., et al. (2013). Overcoming intrinsic multidrug resistance in melanoma by blocking the mitochondrial respiratory chain of slow-cycling JARID1B(high) cells. *Cancer Cell, 23*(6), 811–825.

Ryter, S. W., Alam, J., & Choi, A. M. (2006). Heme oxygenase-1/carbon monoxide: From basic science to therapeutic applications. *Physiological Reviews, 86*(2), 583–650.

Safra, T., Berman, T., Yachnin, A., Bruchim, I., Meirovitz, M., Barak, F., et al. (2013). Weekly topotecan for recurrent ovarian, fallopian tube and primary peritoneal carcinoma: Tolerability and efficacy study—The Israeli experience. *International Journal of Gynecological Cancer, 23*(3), 475–480.

Sarkar, F. H., & Li, Y. (2003). Soy isoflavones and cancer prevention. *Cancer Investigation, 21*(5), 744–757.

Seger, R., & Krebs, E. G. (1995). The MAPK signaling cascade. *The FASEB Journal, 9*(9), 726–735.

Sehouli, J., Stengel, D., Harter, P., Kurzeder, C., Belau, A., Bogenrieder, T., et al. (2011). Topotecan weekly versus conventional 5-day schedule in patients with platinum-resistant ovarian cancer: A randomized multicenter phase II trial of the north-eastern German society of gynecological oncology ovarian cancer study group. *Journal of Clinical Oncology, 29*(2), 242–248.

Sennino, B., & McDonald, D. M. (2012). Controlling escape from angiogenesis inhibitors. *Nature Reviews. Cancer, 12*(10), 699–709.

Setchell, K. D. (1998). Phytoestrogens: The biochemistry, physiology, and implications for human health of soy isoflavones. *The American Journal of Clinical Nutrition, 68*(6 Suppl), 1333S–1346S.

Shamas-Din, A., Kale, J., Leber, B., & Andrews, D. W. (2013). Mechanisms of action of Bcl-2 family proteins. *Cold Spring Harbor Perspectives in Biology, 5*(4), a008714.

Shan, Y., Lambrecht, R. W., Donohue, S. E., & Bonkovsky, H. L. (2006). Role of Bach1 and Nrf2 in up-regulation of the heme oxygenase-1 gene by cobalt protoporphyrin. *The FASEB Journal, 20*(14), 2651–2653.

Siegel, R., Naishadham, D., & Jemal, A. (2013). Cancer statistics, 2013. *CA: A Cancer Journal for Clinicians, 63*(1), 11–30.

Signes, A., & Fernandez-Vizarra, E. (2018). Assembly of mammalian oxidative phosphorylation complexes I-V and supercomplexes. *Essays in Biochemistry, 62*(3), 255–270.

Singh, K. K., Russell, J., Sigala, B., Zhang, Y., Williams, J., & Keshav, K. F. (1999). Mitochondrial DNA determines the cellular response to cancer therapeutic agents. *Oncogene, 18*(48), 6641–6646.

Speicher, L. A., Barone, L., & Tew, K. D. (1992). Combined antimicrotubule activity of estramustine and taxol in human prostatic carcinoma cell lines. *Cancer Research, 52*(16), 4433–4440.

Speicher, L. A., Laing, N., Barone, L. R., Robbins, J. D., Seamon, K. B., & Tew, K. D. (1994). Interaction of an estramustine photoaffinity analogue with cytoskeletal proteins in prostate carcinoma cells. *Molecular Pharmacology, 46*(5), 866–872.

Sun, J., Hoshino, H., Takaku, K., Nakajima, O., Muto, A., Suzuki, H., et al. (2002). Hemoprotein Bach1 regulates enhancer availability of heme oxygenase-1 gene. *The EMBO Journal, 21*(19), 5216–5224.

Tebay, L. E., Robertson, H., Durant, S. T., Vitale, S. R., Penning, T. M., Dinkova-Kostova, A. T., et al. (2015). Mechanisms of activation of the transcription factor Nrf2 by redox stressors, nutrient cues, and energy status and the pathways through which it attenuates degenerative disease. *Free Radical Biology & Medicine, 88*(Pt. B), 108–146.

Viale, A., Pettazzoni, P., Lyssiotis, C. A., Ying, H., Sanchez, N., Marchesini, M., et al. (2014). Oncogene ablation-resistant pancreatic cancer cells depend on mitochondrial function. *Nature, 514*(7524), 628–632.

Vitale, D. C., Piazza, C., Melilli, B., Drago, F., & Salomone, S. (2013). Isoflavones: Estrogenic activity, biological effect and bioavailability. *European Journal of Drug Metabolism and Pharmacokinetics, 38*(1), 15–25.

Xiao, B., Sanders, M. J., Underwood, E., Heath, R., Mayer, F. V., Carmena, D., et al. (2011). Structure of mammalian AMPK and its regulation by ADP. *Nature, 472*(7342), 230–233.

Yang, Y. P., Liang, Z. Q., Gu, Z. L., & Qin, Z. H. (2005). Molecular mechanism and regulation of autophagy. *Acta Pharmacologica Sinica, 26*(12), 1421–1434.

Yano, T. (2002). The energy-transducing NADH: Quinone oxidoreductase, complex I. *Molecular Aspects of Medicine, 23*(5), 345–368.

Yu, O., Jung, W., Shi, J., Croes, R. A., Fader, G. M., McGonigle, B., et al. (2000). Production of the isoflavones genistein and daidzein in non-legume dicot and monocot tissues. *Plant Physiology, 124*(2), 781–794.

Zhang, L., Zhang, J., Ye, Z., Manevich, Y., Ball, L. E., Bethard, J. R., et al. (2019). Heme oxygenase 1 targeting by a cytotoxic isoflavone as a determinant of lung cancer sensitivity. *Cancer Research*. In revision.

CPI Antony Rowe
Eastbourne, UK
March 21, 2019